Astronomy & Astrophysics: Notes, Problems and Solutions

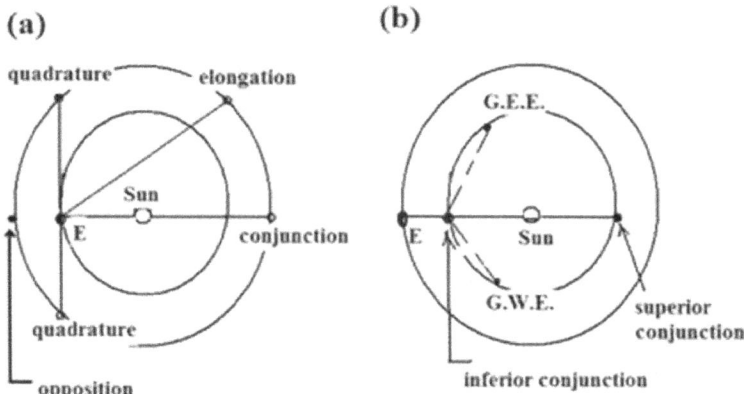

P.A. Stahl
(M. Phil. Physics, B.A. Astronomy)

In Memory of Prof. Heinrich Eichhorn von Wurmb

Table of Contents

Preface   5

Part I: Astronomy

I: Simple Astronomical Timekeeping   7

II. Simple Celestial Sphere Problems   14

III. Stellar Magnitudes - Brightness   22

IV. Astronomical Distance Measurements   31

V. Problems in Sidereal Time   48

VI. Assorted Keplerian Orbit Problems   61

VII. Sidereal –Synodic Periods (Mean Motions)   70

VIII. Computing Relative Distances to Planets   80

IX. Looking at Retrograde Motions   87

X. Basic Celestial Mechanics in 2-Dimensions   106

XI. Binary Star Orbits   127

XII. Introducing Spherical Astronomy   137

Part II: Astrophysics

XIII. Stellar Masses and Luminosities   158

XIV. Stellar Interferometer-Applications   166

XV. Physical Aspects of the Stars 177

XVI. Stellar Evolution Basics 192

XVII. Stellar Atmospheres-Radiative Transfer 204

XVIII. Solar Corona Physics 220

XIX. Galaxies & Density Waves 229

XX. MHD and Solar Physics 239

XXI. Introduction to Special Relativity 253

XXII. Relativity (II): Lorentz Contraction 291

XXIII. The Inertia of Energy 299

Optics Supplement 310

APPENDICES: 338

Selected Solutions to Extra Problems 390

ISBN: 978- 1- 304- 10212- 6

Fourth Edition: Published by Lulu.com Books, (2013). *Copyright, 2012, by Philip A. Stahl. No part of this work may be reproduced, stored in any electronic retrieval system or transmitted by any means without prior permission from the author.*

Preface to First Edition

The present text was probably at least 35 years in the making, preparation. The astronomy and astrophysics content and problems appearing were collected from the time I commenced giving astronomy and astrophysics courses and technical workshops at the Harry Bayley Observatory in Barbados in 1977.

Other content emerged from an extended series of articles I'd written under the banner of 'Mathematics in Astronomy' for *The Journal of the Barbados Astronomical Society* over 1977 – 1990. I had always planned, at some point, to assemble the material into book form but this had to await completion of other projects including an A-Level Physics text and another technical monograph on solar flare plasmas.

The content is of varying difficulty so it will be no surprise that some topics (and their related problems) will be fairly easy to do while others will be much more difficult. I am hoping, however, that readers will at least try a majority of the problems and – if push comes to shove – maybe approach them in a group setting.

In the latter case, the book is also intended to provide stimulus (and workshop) material for amateur astronomical societies. Most of the problems don't require any calculus or high order math, just some background in algebra and geometry.

The sections on spherical geometry and solar physics (including basic radiative transfer through the solar atmosphere and corona), meanwhile, will be most useful for more advanced groups, or individuals who relish a more difficult math arena. For these groups I've also provided (at the end of Appendix II: 'Mathematical Tools') an extensive 'Primer' on differential equations which are really the "skeleton key" to doing any advanced physics, astrophysics. A brief introduction to partial differential equations (the mechanics of solving them, not underlying theory) is also contained supplied with problems.

The chapters on special relativity have been included since this content is so central to modern astrophysics. It is hoped most readers will try at least some of the problems, and again, a group setting is encouraged for the more difficult ones.

I have also included an extensive *Optics Supplement* since optical instruments, their design as well as limitations (e.g. inherent in resolving power) are an integral part of astronomy.

The Appendix also includes sections on physical constants, astrophysical and astronomic data, a primer on nuclear fusion reactions (the basis of stellar energy) and a serviceable glossary.

I hope all astronomy enthusiasts as well as students, interested teachers will find in this volume at least a few things to whet their appetites and also maybe provide hours of learning and enjoyment – which need not be mutually exclusive!

- PHILIP A. STAHL (Colorado Springs, April, 18, 2012)

I. Simple Astronomical Time Keeping

A "*time zone*" is defined by taking the 360 degrees through which Earth rotates in one day, and dividing it by 24, since it requires 24 hours to make one revolution. Thus, one standard time zone would be generated via (360 deg/ 24 hr) = 15 deg/h or 15 degrees of longitude per hour - so be 15 degrees of longitude in expanse. Thus, time zones (calibrated per HOUR) were marked out by LONGITUDE differences.

Time zones don't mean anything until referenced or calibrated to a fixed position-location, and that is the Greenwich Meridian, defined as 0 degrees longitude. All longitudes west of Greenwich mark time earlier – and all longitudes east of Greenwich mark times later. Thus, Berlin will always have a time later than London, and London will have a time later than New Orleans, just as Barbados will always have a time earlier than London and later than Miami.

The time difference is referenced to longitude difference for the central meridians. For example, if London is at approximately 0 degrees longitude, and New Orleans is at 90 degrees west longitude, then New Orleans is earlier than London by (90 deg/ 15 deg/h) = 6 hours. If the time in London is noon local mean time, then it is 6 a.m. in New Orleans.

In order to solve the problem of different local mean times, Greenwich Mean Time or GMT was developed, so people could compare the same clock times around the world. GMT is based on a 24 hour clock defined at the Greenwich Meridian. So, for example, if one is listening to the BBC from New

Orleans and the time given is 13h 30 m GMT, then that means it's 1.30 p.m. LMT in London. Since New Orleans is 6 hours earlier, than that means it's 7.30 a.m. local mean time in New Orleans.

Thus, knowing GMT, one can always work out the time at one's location if one knows the longitude difference relative to Greenwich. (Note for the purposes here, I 'm taking London as having the same longitude as Greenwich. It's actually off by a few thousand feet but negligible in terms of computations.)

Apparent solar time, meanwhile, is erratic because it's based literally on sundial time, and what's called the equation of time (E.T.) (See Figure 1)

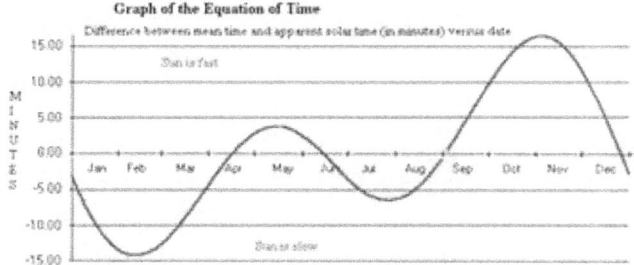

Fig. 1: Graph for the Equation of Time used to obtain apparent solar time.

More useful was the construction of a simple shadow stick such as shown in the diagram below, and simply using it to make apparent solar time measurements. (See Fig. 2 below).

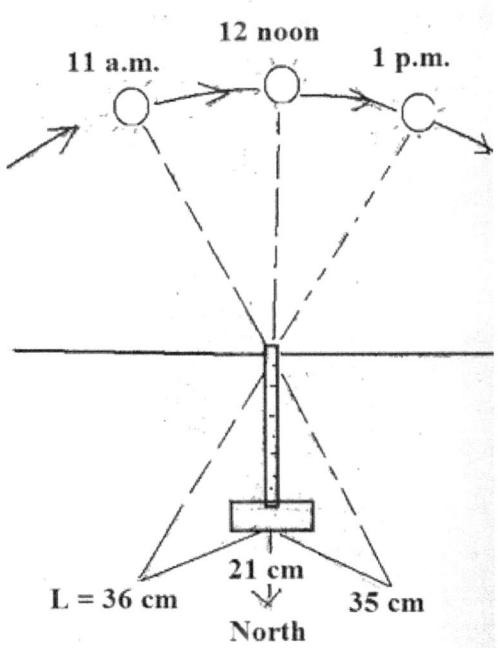

Fig. 2: Shadow stick measurements from Barbados

This simple device was designed by a fifth form student at the Garrison Secondary school, who actually used it to carry out a detailed analysis which was published in *The Journal of the Barbados Astronomical Society*. His measurements of the lengths of the shadow stick are shown for March 21, 1979.

We know that the height (H) of an object placed in direct sunlight is related to its minimum shadow length (L_s) by:

$\tan(a) = H/L_s$

where (a) is the altitude of the Sun. So if H = 100 cm and L_s = 21 cm, then:

$\tan(a)$ = 100 cm / 21 cm = 4.76

And a = arc tan (4.76) = 78.°1

In fact, the actual value for Barbados for the given date should have been 77.° 0 or the zenith distance of the Sun equal to the latitude. This is the precise measurement that would denote local noon apparent solar time.

Exact local mean time for any given longitude is computed via a slight adjustment to standard time. For example, if Barbados actual longitude is 59 degrees 30' minutes W then the local mean time requires a slight adjustment equal to the time difference corresponding to 30' of angular difference, since the meridian referencing Atlantic Standard time (A.S.T.) is 60 deg W and (60 deg W - 59 deg 30' W) = 30'. This is half of a degree, and students in CXC astronomy were shown how to work out that every degree of rotation made by the Earth is equivalent to 4 minutes of time.

Since 15 deg = 60 minutes (1 hour), then 1 degree = 60 mins/15 = 4 minutes. Similarly 30' corresponds to 2 minutes of time. So if A.S.T. (Atlantic Standard Time) at 60 deg W is 2 p.m. then the local mean time for Barbados' specific longitude is: 2 p.m. − 2 mins. = 1: 58 p.m. L.M.T. It was only after much practice that both Caribbean students and teachers became comfortable with these sort of time conversions!

Problems:

1) Study the meridian diagram shown below:

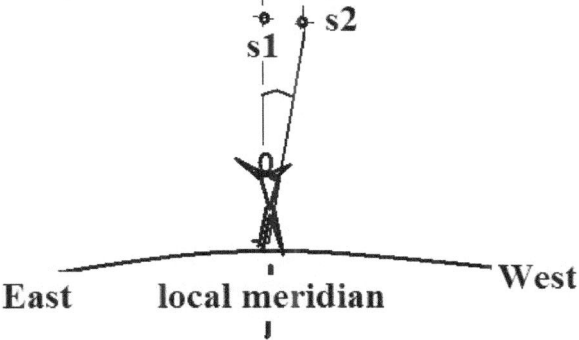

Let the RA = 12 h 30 m for the star at s1 (on the meridian). Position s2 shows the star's position relative to the observer's meridian one day later. Assume it is 7.34 p.m. local mean time for the observer, who wishes to see Antares (RA 16h 34 m).

a) Find the sidereal times and hour angles on each for the dates – for which the observer has s1, then s2.

b) How many hours will the observer have to wait to see Antares on his meridian given condition s2?

c) Find the local sidereal time (LST) for the situation s2 and also find the hour angle of Antares at this time.

2) A student is asked to compute when the star Regulus (RA ≈ 10h 00 m) will transit on his meridian for April 20th. Assume he's asked for *the local mean*

time of transit. How can he work it out? (Assume the student is in Barbados, with longitude: 59° 30' W.

Solutions:

1) a) If the RA = 12 h 30 m for the star at s1 (on the meridian) then the sidereal time at the location is 12 h 30 m. This is also a measure of the time passed since the vernal equinox crossed the observer's meridian.
This, of course, also enables the sidereal time taken the next day (for the same local mean time) to be deduced. Since the hour angle is now equal to the angular equivalent of 4 minutes of time (1°) then the hour angle at s2 = 1°. Or, in other words: the hour angle **(HA) = RA of meridian − RA of object**. We can actually check this easily from the diagram, comparing positions s1, s2 on successive dates.

We already know the RA of the star is 12 h 30 m, associated with position s1. But on the successive day, the star has now moved to position s2, and the meridian RA must therefore be 12 h 34 m. (To account for the difference). Then:

HA = RA of meridian − RA of object

= 12h 34 m − 12h 30 m = 4m, or again, 1° angular measure.

(b) The observer wishes to see Antares with RA 16h 34 m, for the configuration shown (i.e. for ST 12h 34 m) and it is 7.34 p.m. local mean time. Then since the RA is **4h larger than his sidereal time**, then he will need to wait 4 hours for it to reach his meridian, or a local mean time of: 7.34 pm. + 4h 00m = 11 .34 p.m.

(c) From this, one sees the local sidereal time (LST) at one's location is just:

LST = HA + RA of object

Thus if LST = 12 h 34m and RA = 16h 34 m then:

HA = LST – RA of object – 12h 34m – 16h 34m = -4h

The minus sign means the object *is east of the meridian*, and since 15 degrees = 1 hour of time, the actual value for HA = - 60 °.

2) The main impediment for the student here is to obtain *the Sun's Right Ascension*. He can do this without too much fanfare if he recalls that a month earlier, March 20, the Sun was approximately at the vernal equinox and therefore its RA = 0h. April 20 is one month later, translating to an advance of 4 minutes per day for 30 days or: 4 min/day x 30 days = 120 minutes.

That is, two hours of total time difference. So the Sun's approximate RA = 2h 00m. Then the local mean time for transit (LTT) is:

LTT = 10h 00m – 2h 00m = 8h 00m

In other words, Regulus transits 8 hours *after the Sun* (which always transits at noon LMT) so the local mean time of transit must be: 8 p.m. + 2 minutes = 8.02 p.m., say for an observer in Barbados, to account for the 30' difference in longitude between 59° 30' and 60° (referenced to Atlantic Standard Time

II: *Simple Celestial Sphere Problems*

Problems to do with the celestial sphere (Fig. 1) and locating object positions in the sky also formed a major part of the work in the CXC astronomy syllabus.

FIG. 1

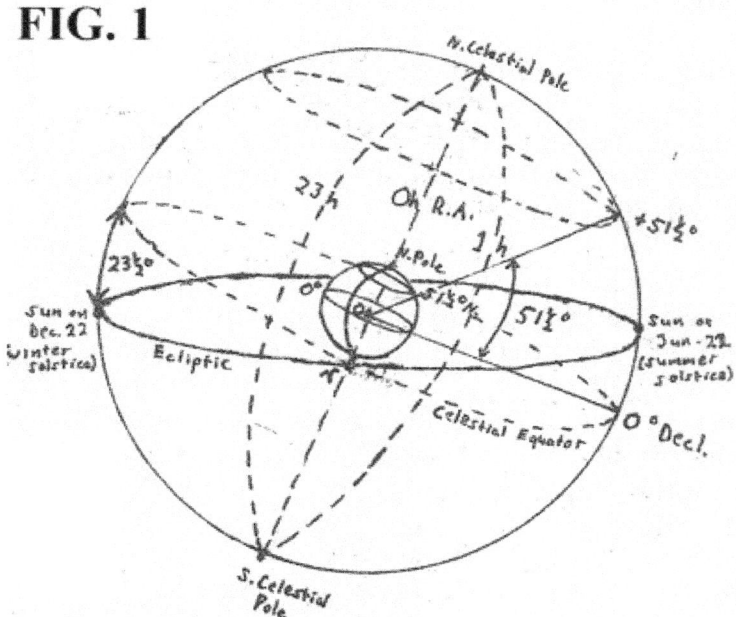

The diagram of Fig. 1 is from an article I wrote in the 2nd Volume (issue No.4) of *The Journal of the B.A.S.* to do with celestial coordinate systems, and transforming from one to the other. The key aspect is the generation via projection of the larger coordinate system (from the Earth's latitude and longitude system), from which advanced students could make all needed conversions, say from the celestial system (R.A., Declination) to the horizontal (altitude and azimuth).

For the more pedestrian students it was enough to teach them the use of declination diagrams (Fig. 2) to work out simple problems.

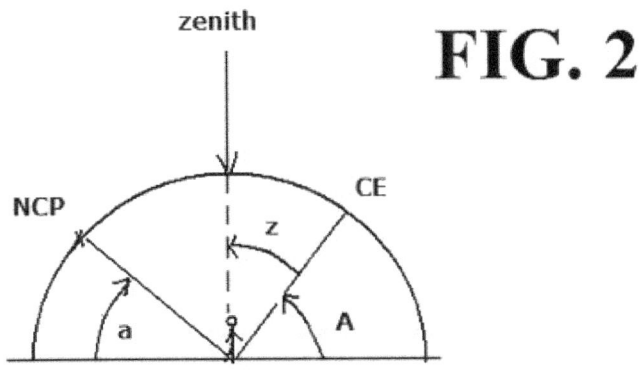

FIG. 2

NCP = North Celestial Pole

a = alt. of NCP = lat.

z = zenith distance = a

A = alt. at meridian transit

CE= Celestial equator (0 deg Declination)

The basic elements are evident in the diagram. In effect, rather than having to work in the realm of difficult spherical geometry, the student was asked merely to get comfortable with working using plane geometry!

Even so, it was a major challenge for most students, and teachers! Specialist teacher trainers and teachers themselves most often found that while the general principles, plane geometry were fairly clear, most of the applications weren't. Consider this example which seeks to find the star's declination.

The altitude of a star as it transits your meridian is found to be 45 deg along a vertical circle at azimuth 180 deg, the south point. Find the declination of the star.

Since this was designed for students at latitude 13 degrees north, the key to the solution rested on the recognition that z, the zenith distance was negative. From the geometry of Fig. 2 one saw that:

$90° = z + a$ or $z = 90° - a = 90° - 45° = 45°$

But since we require: $z = \varphi$ (latitude) = 13°

Then z must have a negative value, or: (-45°), since:

$\delta = z + \varphi = (-45°) + 13° = -32°$

This makes sense, since by examining the right side of Fig. 2., the zenith distance z, plus altitude (a) must equal 90 degrees and we know CE (celestial equator) defines 0 degrees declination, then a star's altitude of a = 45° shows it to be SOUTH of CE. How much? Ans. $90° - 45° = 45°$.

But, this is still 32° south of CE, and hence must be *negative* in value. (Remember CE is only 13° from the zenith point in Barbados). Of course, most zenith diagrams in tests were deliberately drawn not to scale, in order to make sure students grasp the principles and really attend to the geometry.

As the activity progressed, most CXC teachers found they had to go slow, keep the examples very basic and avoid synthesis with other activity objectives.

Some of the more advanced students (e.g. S.T.A.R. members) needed to see what more could be done with azimuth (trickier than altitude- measured along the observer's horizon) and they were provided extra worksheets, e.g. in seminars. One of the favorite sub-activities was to find the azimuth of the Sun for non-Caribbean locations, such as London, say for the rising or setting times at winter and summer solstice.

In general:

$\cos(A) = \sin(\delta)/\cos(\varphi)$

where A is the azimuth of the Sun, δ denotes its declination, and φ is the observer's latitude. (Note that δ may be obtained from a table but can also be estimated from the equinox/solstice positions, i.e. $\pm 23\frac{1}{2}°$ at solstices, 0 degrees at equinoxes.

Example: $\varphi = 51.5$ degrees N, for London. Now, for the December (winter) Solstice the Sun is directly over the Tropic of Capricorn ($\varphi = 23.5$ S) therefore we do know its declination is $-23.°5$. We have for the Sun's azimuth at sunrise on Dec. 21:

$\cos(A) = \sin(-23.°5)/\cos(51.°5)$

which gives approximately, $130°$.

Where is this on our directional reference circle for azimuth? We know that 180 degrees is *due South* so that this must be:

40 degrees SOUTH of due East. (90° + 40° = 130°)

Now, on *the longest day of the year* (say June 21), the Sun is over the Tropic of Cancer at 23.5 N latitude, so the Sun's declination is + 23.°5 . Then the azimuth for that date is:

cos (A) = sin (23.°5)/ cos (51.°5)

And A = 50 °

This puts the Sun's rising position North of due E. or specifically 40 degrees *North of due East*. Based on the preceding results, between the shortest and longest day of the year, one would see the Sun move from south of due east to north of due east, by the amount of degrees difference indicated.

This also discloses that on the longest day one can't observe the Sun at true geographical East! But rather forty degrees North of that position at rise time. So on the shortest day (June 21), one will be seeing the Sun rise 40 degrees south of East, or approximately south to south-east.

Caribbean students thereby became aware that while the cardinal directions of the compass agree with the azimuth directions, starting from 0° and separated by 90°, this didn't apply to other much northerly locations.

Determining observational limits was also easy to do with the use of declination diagrams such as in Fig. 2. Here are two easy problems for readers to try their hands at:

1) Draw a declination diagram for London (lat. 51.5 N). What would be the most southerly star visible by declination? Which stars would be circumpolar? What declination parallel would pass through your zenith?

2) What is the maximum altitude which would be attained by Alphecca (declination +26 ° 50') at Barbados (latitude 13 deg N).

What would the meridian zenith distance of Alphecca be?

Solutions:

(1) The declination diagram constructed for London, latitude 51.°5 N, is shown below:

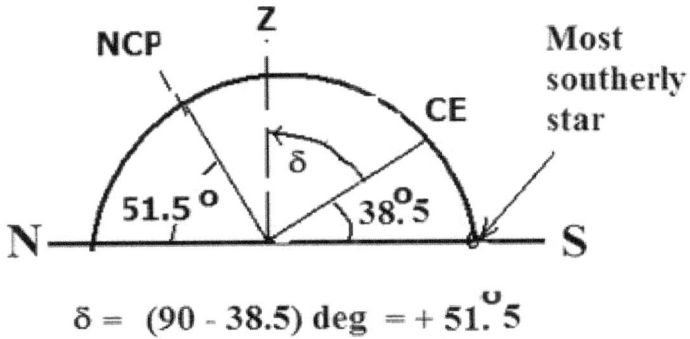

δ = (90 - 38.5) deg = + 51.°5

Decl. Diagram for London

The most southerly star by declination is marked on the southern horizon. Since CE (the celestial equator) and NCP (*the north celestial pole*) must be 90 degrees apart, and CE defines the circle for 0 degrees declination, then the most southerly declination is: - 38.°5 or 38.5 degrees south of the CE.

The circumpolar stars would be those observed from London which over time describe a circle around the NCP. Thus the condition for circumpolarity would be:

51°.5 ≤ δ ≤ 0

where δ denotes declination. The declination of all the stars that would pass through the Londoner's zenith is given by the angle delta extending from the CE (at 0 degrees) to z.

This is: (90° - 38.°5) = + 51.°5

Hence, all objects with a declination of +51.°5

(2) This solution requires a declination diagram for Barbados (lat. 13 deg N) which is shown below:.

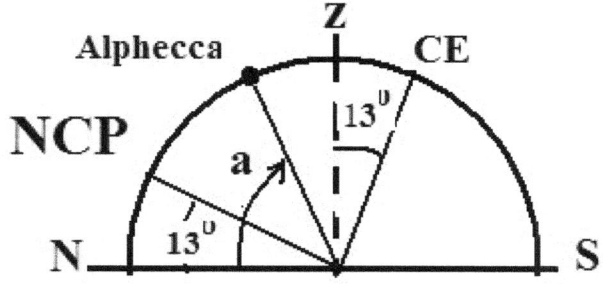

The declination diagram for Barbados (lat. 13 deg N)

The star of interest is Alphecca: declination +26° 50'. The maximum altitude a, attained by Alphecca, is easily computed from:

a = (90° - z)

where z denotes the zenith distance (or distance from the observer's zenith).

We know that all objects with declination $\delta = 13°$ pass through the zenith, therefore the zenith distance z for an object of declination $+26° 50'$ must be:

$z = (26° 50' - 13°) = 13° 50'$.

Note this is also the *'meridian zenith distance'*.

Therefore, the maximum altitude, a:

$a = (90° - 13.° 50') = 76° 10'$

Other Problems:

1) Jacksonville, Florida is located at approximately latitude $30° 20'$ N.

a) Sketch a declination diagram for this location and carefully identify the position of the North Celestial Pole and the approximate altitude of the Pole star.

b) If the star Procyon is visible and has a declination of approximately $\delta = +5°$ then find its maximum altitude from this location and its zenith distance, z.

2) Find the *azimuth* of Procyon for the same location given in (1), and estimate the date if the Sun's declination is known to be $+7° 20'$.

3) How would the values obtained in (2) for Procyon (and Jacksonville) change for a location in Barbados?

III: Stellar Magnitudes- Brightness

One of the most fundamental astronomy areas, which also gave teachers and students doing CXC Astronomy fits, concerned the matter of relative and absolute brightness and stellar magnitudes. (Specifically, the stellar magnitude scale). Of course, it's always best to begin such a topic with prosaic examples.

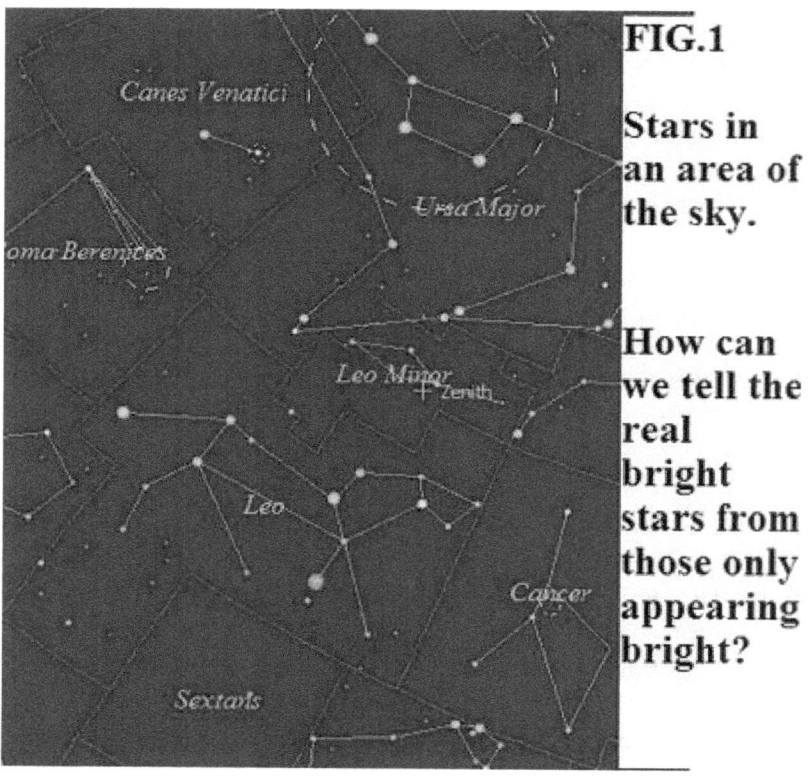

FIG.1

Stars in an area of the sky.

How can we tell the real bright stars from those only appearing bright?

I tended to use the simple, everyday example of comparing two light bulbs. Thus, if I have a 100 watt bulb and place it at 1 m, then what would its relative brightness be when placed *twice as far away*? By the inverse square law for light (the intensity of a light source decreases as the inverse square of its distance):

$(d_1/d_2)^2 = B_2/B_1$

Where in this case, $d_1 = 1$ m, $B_1 = 100$ w and $d_2 = 2 d_1 = 2$m. Then, it's obvious one needs to find B_2.

$B_2 = B_1 (d_1/d_2)^2$

But $(d_1/d_2)^2 = (1/2)^2 = 1/4$

Therefore: $B_2 = 100$ w $(1/4) = 25$ watts

The inverse nature of the proportion appeared to trip up many students. For many others, it was the squaring of the distance term.

In the astronomical context, the student ought to be able to use the same method in applying the inverse square law to stars. But actually doing it proved to be the bugbear! For example:

Beta Crucis has an apparent magnitude of +0.8 from 261 light years. Find its absolute magnitude.

The student is first given the basis for the absolute magnitude as that apparent magnitude defined at a distance of 10 parsecs, and if there are 3.26 light years per parsec this means, 10 pc (3.26 ly/pc) = 32.6 light years. Then, for Beta Crucis:

$(d_1/d_2) = 261$ ly$/ 32.6$ ly $= 8$

Then:

$(d_1/d_2)^2 = (1/8)^2 = 1/64$

So:

$B_2/B_1 = 1/64$

But in this case the object is being *brought closer*, so we want to find B_1! Thus, $B_1 = 64 B_2$. In other words, the star's brightness at 32.6 light years is roughly **64 times more than its brightness at 261 light years** which converts to a magnitude difference of: 4.5. (See Fig. 2 below for a stellar magnitude scale with the Sun at the extreme left or brightest end, with an apparent magnitude of (-27).

FIG. 2

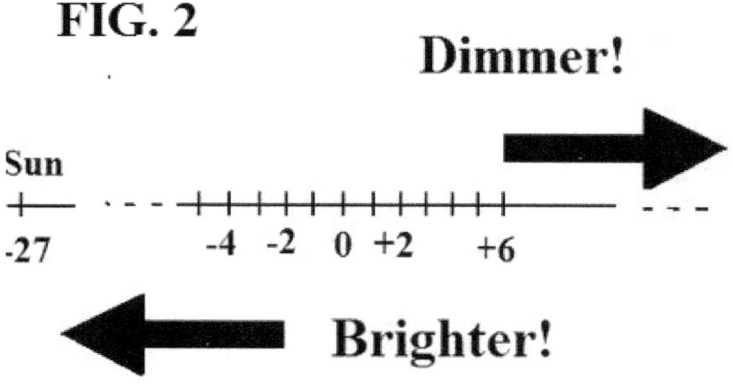

So the new magnitude would be: $(0.8 - 4.5) = -3.7$.

That is, we locate +0.8 on the scale, then march off 4.5 units on it in direction of increased brightness.

The magnitude scale and magnitude difference introduces a logarithmic scale on top of the ratios.

The layout of the scale is arithmetical, very much like the algebraic number line, but when related to brightness one obtains *ratios* not differences. And these ratios are based on a logarithmic relation. For example, the apparent magnitude of the Sun makes it apparently the brightest object in the sky. But this is only because it is at 1 A.U. distance. IF "moved" to 32.6 light years (the standard candle distance to obtain absolute magnitude) then one obtains +4.8 magnitude.

Thus, its real magnitude in terms of the absolute standard (absolute magnitude) makes it somewhat dim.

The basic gist of the scale is that every 5 increments (UNITS) in magnitude difference translates into 100 times difference in brightness, because each succeeding magnitude is different from the earlier one by 2.512 times, since: $(2.515)^5 = 100$. Thus a star of +1 magnitude is 100 times brighter than a star of +6 magnitude on the scale, since $(6 - 1) = 5$.

Problems:

(1) (a) Using the brightness-magnitude scale in Fig. 2 as a basis construct a simple stellar magnitude scale ranging from (-10) to (10).

i) Sirius has a magnitude of (-1.4), mark it on the scale and label it.

ii) Castor has a magnitude of (+1.6) mark its place on

the scale and label.

iii) Sigma Draconis has a magnitude of (+4.7), mark and label it.

iv) Zosma has a magnitude of (+2.55), mark in and label it.

(b) Which is brighter, Sirius or Castor, and by how many times? Which is brighter, Zosma or Sigma Draconis and by how many times? A new star Alpha Stellaris is discovered which is found to be 15.6 times brighter than Sirius. Locate its approximate position on the scale and label it. Estimate how many times brighter this new star is than Sigma Draconis.

(2) One of the stars shown on the star map of Fig. 1 is Regulus in the constellation Leo. It has an apparent magnitude of (+1.35) and absolute magnitude (-0.6). Using this information only, estimate its actual distance from Earth.

(3) The Sun's absolute magnitude is +4.8. How much farther would it have to be from us to appear with that magnitude in the night sky? (The Sun's mean distance from us currently is 150 million kilometers for which the apparent brightness is (-27).

(4) Two stars in the same constellation are named Alpha and Gamma. Both have the same absolute magnitude. Alpha is at a distance of 32.6 light years and has an apparent magnitude of +0.8. Gamma is at a distance of 261 light years.

a) Using the inverse square law of light, estimate how

many times Gamma is dimmer than Alpha.

b) Based on the stellar magnitude scale, estimate Gamma's apparent magnitude.

Solutions to Problems:

(1) (a) Using the scale in Fig. 2 as a basis construct a simple stellar magnitude scale ranging from (-10) to (10). The resulting diagram is shown:

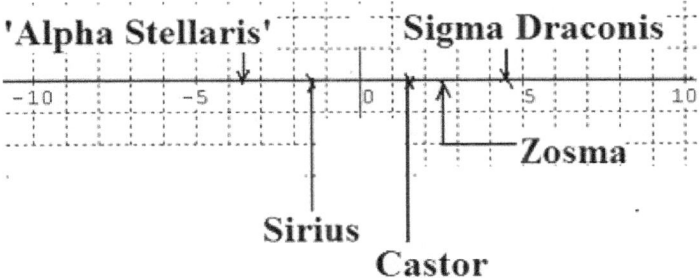

Problem 1: Stellar magnitude scale and star identifications from app. magnitude

Marked down are: Sirius (magnitude of -1.4), Castor (magnitude of +1.6), Sigma Draconis (magnitude of +4.7), and Zosma (magnitude of (+2.55).

Solution (b):

The constructed magnitude scale shown can again be used, with the stars identified by their apparent magnitudes, including 'Alpha Stellaris'.

As to the answers for part (b):

Sirius is at m = (-1.4) and Castor at +1.6 so the **magnitude difference** is:

[1.6 - (-1.4)] = 1.6 + 1.4 = 3.0

so the difference in brightness (brightness ratio of Sirius to Castor) is:

$(2.5)^3$ = 15.6 times

Alpha Stellaris is 15.6 times brighter than Sirius therefore (from the previous working) it will be three units of magnitude brighter, hence will be at:

(-1.4) - 3.0 = -4.4 magnitude

Relative to Sigma Draconis, Alpha Stellaris would be:

+4.7 - (-4.4) = 9.1 magnitudes brighter or

$(2.5)^{9.1}$ = 4,180 times brighter

(2) Regulus' magnitude (by definition of absolute magnitude) would be -0.6 at 10 pc. Therefore, its apparent magnitude being (+1.35) it is actually *further away than 10 pc*. The magnitude difference is:

+1.35 - (-0.6) = +1.35 + 0.6 = 1.95

or a brightness ratio (b/b') = $(2.5)^{1.95}$ = 5.97

but: d'/d = $(b/b')^2$

so d' = $(5.97)^{1/2}$ d = 2.44 d

therefore d' = 2.44 x 10 pc = 24.4 pc (actual distance)

(3) The Sun's absolute magnitude is +4.8. How much farther would it have to be from us to appear with that magnitude in the night sky? (The Sun's mean distance from us currently is 150 million kilometers for which the apparent brightness is (-27).

Solution

First obtain the difference in magnitudes:

(+4.8 - (-27) = 31.8 units of magnitude.

Then the brightness ratio is:

$(2.5)^{31.8} = 4.5 \times 10^{12}$

But the distance is related as the inverse square, so:

$(d/d') = 1/(4.5 \times 10^{12})^{1/2} = 2.12 \times 10^6$

so the Sun would have to be about 2.1 million times farther away (e.g. 3.1×10^{14} AU)

(4) Two stars in the same constellation are named Alpha and Gamma. Both have the same absolute magnitude. Alpha is at a distance of 32.6 light years and has an apparent magnitude of +0.8. Gamma is at a distance of 261 light years.

a) Using the inverse square law of light, estimate how many times Gamma is dimmer than Alpha.

b) Based on the stellar magnitude scale, estimate Gamma's apparent magnitude.

--

Solution

Part (a) just requires taking the distance ratio:

d_2/d_1 = 261 LY/ (32.6 LY) = 8.0

so by the inverse square law, the brightness of Gamma (being eight time more distant) must be 1/64 of Alpha's.

Part (b) is solved by simply applying the logarithmic features of the stellar magnitude scale. Again, I remind readers that every unit magnitude difference (arithmetical) is equal to a brightness ratio of 2.5 times. Thus, to obtain the apparent magnitude of Gamma, we need to solve:

$(2.5)^x$ = 64 or: x (log 2.5) = log (64)

Then: x(0.39) = 1.80 and x = (1.80)/ 0.39 = 4.6

Which means Gamma's apparent magnitude is:

m = (0.8) + 4.6 = +5.4

IV. Astronomical Distance Measurements:

The most basic and intuitive method for measuring a star's distance employs elementary trigonometry. It is called the *"parallax method"* because it uses the phenomenon of "parallax". One can get a simple idea of this by holding an index finger at arm's length- then looking at it with only the right eye open, then only the left. What you'll see as you do this alternatively is the visible shift of your index finger against the given background. Say from one part of your book shelf to another. (A very tiny shift.)

The angle that is subtended between the different background points, with respect to your nose ("vertex" of the resulting triangle), say, is the "parallax angle" Now, let's generalize this to stars, using the diagram below - which I hope comes out reasonably well in the response:

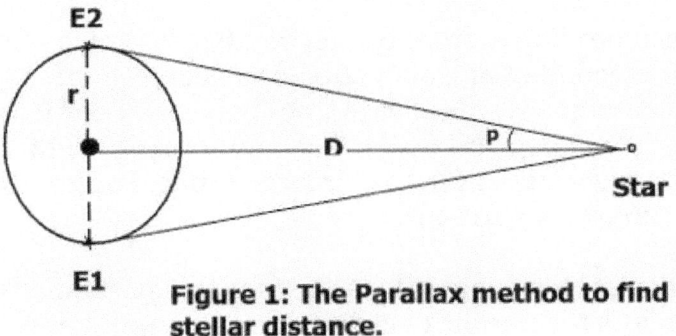

Figure 1: The Parallax method to find stellar distance.

Here, the points E1 and E2 represent Earth at two opposite points of its orbit around the Sun. The star indicated is the one for which the distance(d) is

31

sought. Using photographs of the star taken at the points E1 and E2, we can measure what is called the angle of parallax p. The solution for the distance can be obtained from:

d = r/ tan(p)

that is, equal to the radius (R) of the Earth's orbit, e.g 1 A.U. or astronomical unit = 93 million miles = 1.5 x 10^{11} meters) divided by the tangent of the angle p. (which will be a very small angle). The tangent is a function which denotes a ratio of lengths for a right triangle. In this case, the tangent of the angle p is defined:

tan(p) = opposite/ adjacent = R/d

i.e., the side opposite the angle (R), divided by the side (d = distance to star) adjacent to the angle (d). Using algebra, one then makes d the subject, which means solving for it.

It can be shown, from the above, that for angle p = 1" (one second of arc, or 1/3600 of a degree!) the distance to the star is always 1 pc (parsec). Bear in mind 60' = 1 deg and 60" = 1' or 1 minute of arc) the resulting d = 206, 265 A.U. or astronomical units (206, 265 x 1.5 x 10^{11} m).

The exact term logically chosen for this particular distance is the parallax-second or *parsec* and doing the math (tedious, but possible) shows the distance turns out to be:

1 parsec = 3. 26 light years.

Now, given this relationship, it follows that any proportionate decrease in the angle p allows one to work out the resulting distance! Thus, if p = 0.5" then d = 2 parsecs or 6.52 Ly, if p = 0.25" then d = 4 parsecs or about 13 Ly and so on. Thus it was found that even the nearest stars were light years distant - using the simplest conceptual method available. The first successful parallax measurements - so far as we know - took place around 1838, when Friedrich Bessel (Germany), Thomas Henderson (Cape of Good Hope) and Friedrich Struve (Russia) detected the parallaxes of the stars 61 Cygni, Alpha Centauri and Vega, respectively.

No one knows who "first" used it. Parallaxes have been measured for thousands of stars. However, for barely 700 are the angles (p) large enough (about p = 0.05" or more) to be measured with a precision of 10% or better. (Which means the resulting distances will have the same order of uncertainty).

Most of these measurable stars are within 20 parsecs or around 66 light years. Clearly, other methods are needed to measure larger distances. Among these is the method of Cepheid variables. These are a special type of star for which the brightness changes as its surface swells (expands) then contracts. In 1912, a period-luminosity relationship was discovered for these stars, by Henrietta Leavitt, an astronomer at Harvard College Observatory. Most of these were in the satellite galaxies (to our own) known as "the Small" and "the Large" Magellanic Clouds. Leavitt's now famous "law" for period-luminosity (relating the time to vary to intrinsic brightness) and its applications is found in many astronomy books.

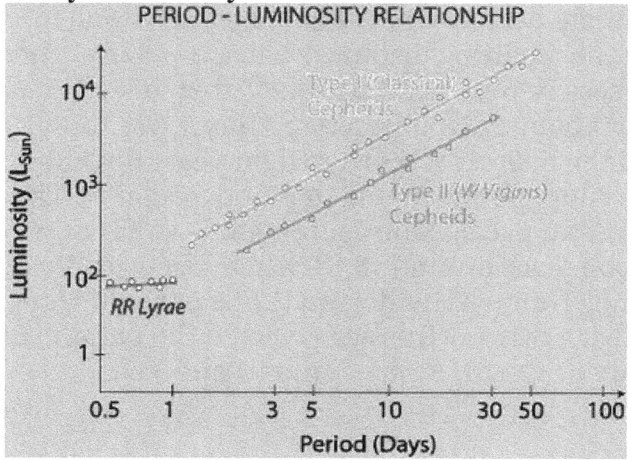

Employing this law enables us to use a kind of a standard candle to measure distance, based on the principle of the inverse square law for light (see previous chapter for details).

What one actually sees is *the apparent magnitude* of the stars plotted against the logarithm of their period (in days). Thus, there is a relation between the apparent magnitude of the star and period of its light curve. Now, since there is a direct relation between apparent and absolute magnitude (the absolute magnitude of a star is just its apparent magnitude from the same standard distance of 10 parsecs or 32.6 light years) there must also be a direct relationship between a (Cepheid) star's period and absolute magnitude.

Thus, if a given set of Cepheids have absolute magnitudes assigned to them then the relationship between their periods and absolute magnitudes acts as a distance measuring indicator. Entering into this is what is called the "distance modulus" (m - M) given as the difference between apparent magnitude (m) and absolute magnitude (M):

$(m - M) = 5 \log_{10}(r) - 5$

Here, r is the distance to be found. For example, a star is found to have: m = +4.5, and M = +6.0 (meaning it is actually *brighter in apparent magnitude* than in reality).

We have:

$(4.5 - 6.0) = 5 \log_{10}(r) - 5 (-1.5) + 5 = 5 \log_{10}(r)\ 3.5$

$= 5 \log_{10}(r)$

then: $\log_{10}(r) = 3.5/ 5 = 0.70$

taking anti-logs:

r = 5.0 (parsecs)

Which is actually for the star 40 Eridani.

Now, in the case of Cepheids - to employ the P-L relation for any star, we need to first find a 'zero point' applicable (this is not easy by any means, and I don't intend to go into the technical details!) By choosing the correct "law" appropriate to the type of variable star observed (e.g. 3-day period, 10-day period, 30-day period etc.) a value can be inserted for the absolute magnitude M.

Then we can again apply the distance modulus equation, as I demonstrated above, since we now have values for both m and M. Note that the Cepheid method is good for many stars in nearby galaxies (like the Magellanic clouds) or in star clusters. However, one must realize problems can arise - for example many of these Cepheids are in dusty regions of the Milky Way with some light lost by absorption - making them appear less bright than they'd normally be at that distance. Hence, if such anomalous Cepheids are used for finding a distance they'd give a skewed, erroneous result. At much larger distances, say for galaxy clusters - we make use of the so-called "*Hubble relation*" or law.

That is, Edwin Hubble first discovered that the galaxies are speeding away from us with velocities proportional to their red shifts. (This is in reference to the shift of observed, known spectral lines toward the red or longer wavelength region disclosing movement *away* from an observer. This observation for multiple distant objects has revealed the "expansion" of the universe).

The red shift is given by:

$z = v/c$

where c is the velocity of light (300,000 km/s)

For example, if the H-alpha spectral line is found to redshift by 20% from its normal position (at 6563 Å where 1 Å = 10^{-8} cm) we have:

0.2 = v/c or v = 0.2 c

In other words, its velocity of recession is two-tenths the speed of light, or 0.2 (300,000 km/s) = 60,000 km/s

The distance can then be found from the Hubble law:

v = cz = HD

where H is Hubble's constant, and D is the distance of the galaxy cluster or other object (e.g. quasar)

We have H = 100 km/ sec (Mpc $^{-1}$)according to recent measures, so that using the example of the recessional velocity, v above:

D = v/ H = 60,000 km/s / [100 km/ sec (Mpc $^{-1}$)]D

= 600 Mpc (Mega-parsecs)

That is, 600 x (10^6) parsecs = 6 x 10^8 parsecs

or 6 x 3.26 x 10^8 light years = 1.95 x 10^9 light years

Or, about two billion light years.

The preceding examples, starting with the simple parallaxes, shows us that distance measurement (and associated techniques) in astronomy are by no means straightforward. This is also possibly why it is difficult to get satisfactory answers, since there are so many different methods appropriate to differing distance scales. In addition, to understand each method, both mathematics, and a certain amount of physics (e.g.

inverse square law for light, spectral line shifts etc.) enter at each point. Therefore, appreciation of the distance methods and their appropriate use, really depends on how much of the physics - and math - one is able to incorporate in order to ascertain for himself how it works.

Sample problems:

(1) A star has an apparent magnitude m = +3.5. Find its absolute magnitude M if its distance = 10 pc.

Solution:

The absolute magnitude is defined as the magnitude the star would have at a distance of 10 parsecs. Since the star is already at that specified distance, then M = +3.5, or the same as its apparent magnitude.

(2) Barnard's star has an absolute magnitude of +13.2 and an apparent magnitude m = +9.5. Find its distance in LIGHT YEARS.

Solution:

We may use the parsec form of the distance modulus:

$(m - M) = 5 \log (1/p) - 5$

$(9.5 - 13.2) = 5 \log(1/p) - 5$

$-3.7 = 5 \log (1/p) - 5$

$5 \log (1/p) = (5 - 3.7) = 1.3$

log (1/p) = (1.3)/5 = 0.26

Taking anti-logs:

1/p = D = 1.81 pc

But 1 pc = 3.26 Ly, so D = (1.81 pc)(3.26 Ly/pc) = 5.9 Ly

Extra Problems for the motivated reader:

(1) A star with an apparent magnitude of +1.0 is located at a distance of 40 pc. What is its absolute magnitude M? What is its *distance modulus*?

(2) Find the apparent magnitude of the star Vega if its absolute magnitude is +0.3 and its parallax angle p = 0."123.

(3) The star Pollux in the constellation Gemini has a parallax angle p = 0."093. Find its distance in light years.

(4) Altair in the constellation Cygnus is 16.4 LY distant with an apparent magnitude m = +0.8 and absolute magnitude M = +2.3. Verify its absolute magnitude using the *inverse square law for light*.

Selected Solutions:

1) We know m = +1, and D = 40 pc.

Then, from the distance modulus, the absolute

magnitude M:

M = m + 5 - 5 log D

M = +1 + 5 - 5 log 40

or: M = +6 - 5 log 40

log 40 = 1.602

So:

M = +6 = 5 (1.602) = +6 - 8.01 = -2.01

Then: the *distance modulus* is:

(m - M) = +1 - (-2.01) = 3.01

2) We have: M = +0.30 and p = 0."0123

The apparent magnitude of Vega may be determined from the distance modulus in the form:

(m - M) = 5 - 5 log p

So: m = M - 5 - 5 log p

or:

m = +0.30 - 5 - 5 log (0.123)

m = -4.70 - 5 (-0.91) = -4.70 + 4.55 = - 0.15

3) Since the trigonometric parallax p = 0."093 we can find the distance in parsecs directly, since:

D = 1/p

Thus, D = 1/ (0.093) = 10.75 pc

But we need D in light years, and know 1 pc = 3.26 Ly, so:

D = 10.75 pc (3.26 pc/Ly) = 35 Ly, approx.

4) Altair has an apparent magnitude m = +0.8 at 16.4 Ly, but 16.4 pc = 5 pc

I.e. 5 pc x 3.26 Ly/pc = 16.4 Ly

Absolute magnitude M is based on a standard distance of 10 pc.

By the inverse square law for light:

$(D/D')^2 = (B'/ B)$

Where B'/B denotes the brightness ratio.

$D/D' = 1/2$

So: $B' = (1/2)^2 B = 1/4\ B$

i.e. the brightness at 10 pc must be decreased by a factor 4.

Then:

$(2.512)^n = 4$

n log (2.512) = log 4

n (0.4) = 0.60

n = 0.60/0.40 = 1.5

In other words, to get M one must add (1.5) to m. Hence:

M = (+0.8) + 1.5 = +2.3

Problem:

On the graph below showing the *Period-Luminosity law* two Cepheids, A and B, are shown:

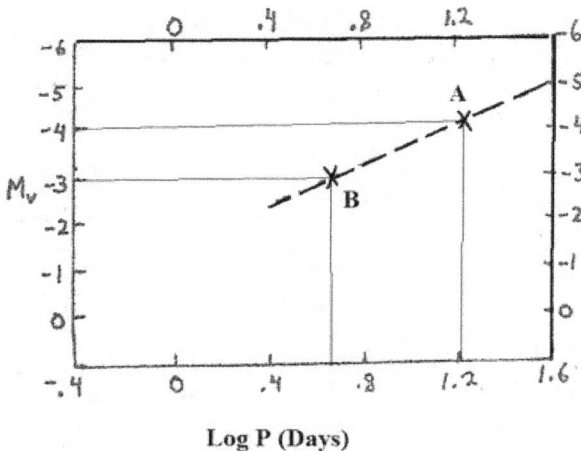

Log P (Days)

a) Compare the periods of the two stars, in days.

b) How much brighter is the longer period star than the other?

Solutions.

a) The periods can be read off along the horizontal axis. For Cepheid A we have Log P = 1.2, so P = 15.8 days (e.g. antilog (base 10) of 1.2 = 15.8). Similarly, for Cepheid B we have Log P = 0.65 (est.) so P = 4.4 days.

b) The *longer period* Cepheid (A) is 2.512 times brighter than the shorter period one, B, since there is 1 magnitude difference,

$M_v(A) - M_v(B) = [(-4) - (-3)] = -1$

and each magnitude is different from one that's a unit brighter, by 2.512 times, since: $(2.512)^5 = 100$.

Other Problems:

1) For the example problem given, if the distance to Cepheid A = 10 pc, find the distance to Cepheid B.

2) One Cepheid is found to have a period of 6.32 days and a mean apparent magnitude of +4.5. Another Cepheid in the same constellation is found to have a period of 15.9 days and a mean apparent magnitude of +7. If the P-L relationship for both has an uncertainty of $\pm 0.35^m$, find the approximate distance for each one.

3) Two Cepheids, Alpha and Beta are observed to have the same period of 10 days. at maximum brightness A has an apparent magnitude of +3.0 and B has an apparent magnitude of +8.0. If the distance of A

(associated with a cluster) is known to be 60 pc, how far away is B?

FIGURE 2

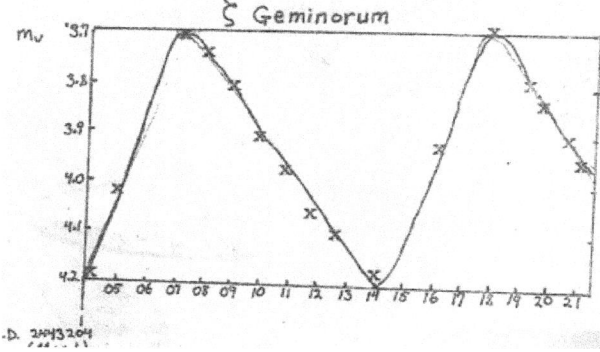

4) A CXC astronomy student plots a light curve for the Cepheid Zeta Geminorum as shown above. Using this curve and the P-L graph, estimate the brightness of this Cepheid in terms of absolute visual magnitude.

Solutions:

We already know Cepheid (A) is 2.512 times brighter than the shorter period one, B. Thus, B(A) = 2.512 B(B).

We also know: d(A) = 10 pc

From the inverse square law for light:

$(d_1/d_2)^2 = B_2/ B_1$

or, in this case, letting $d_2 = d(A)$, so $B_2 = B(A)$:

$[d(B)/d(A)]^2 = 2.512$

and:

$[d(B)/10 \text{ pc}]^2 = 2.512$

$[d(B)/10 \text{ pc}] = [2.512]^{1/2} = 1.58$

Then: $d(B) = 1.58 (10 \text{ pc}) = 15.8$ pc or 51.5 Ly

2) One Cepheid is found to have a period of 6.32 days and a mean apparent magnitude of +4.5. Another Cepheid in the same constellation is found to have a period of 15.9 days and a mean apparent magnitude of +7. If the P-L relationship for both has an uncertainty of $\pm 0.35^m$, find the approximate distance for each one.

Solution

For the Cepheid with Period(P) = 6.32 days, $\log P = \log(6.32) = 0.8$

For the Cepheid with Period(P) = 15.9 days, $\log P = \log(15.9) = 1.2$

Using the P-L graph (see last blog) the intrinsic brightness associated with a period of $\log P = 0.8$ is $M_v = -3.2$, and for $\log P = 1.2$ it is (-4). Assuming M_v = abs. magnitude, and using the distance modulus equation:

$(m - M) = 5 \log r - 5$, we have:

for Cepheid (1):

$\log r_1 = (m - M + 5)/5 = [\{+4.5 - (-3.2) + 5]/5 = 12.7/5 = 2.54$

and: antilog$_{10}$ (2.54) = 346.7 so r_1 = 346.7 pc

for Cepheid (2):

$\log r_2 = [+7 - (-4)+5]/5 = 16/5 = 3.2$

and antilog$_{10}$(3.2) = 1584.9

so: r_2 = 1584.9 pc

Now, let r_1 and r_2 have *probable errors of* e_{r_1} *and* e_{r_2}, respectively. Then:

$e_{r_1}/r_1 = -0.46\ (\delta M)$ and $e_{r_1}/r_1 = -0.46\ (\delta M)$

Finally:

$e_{r_1} = (346.7\text{ pc})(0.46)(\pm 0.35) = \pm 55.6$ pc

and

$e_{r_2} = (1584.9\text{ pc})(0.46)(\pm 0.35) = \pm 255.2$ pc

3) Two Cepheids, *Alpha* and *Beta* are observed to have the same period of 10 days. At maximum brightness A has an apparent magnitude of +3.0 and B has an apparent magnitude of +8.0. If the distance of A (associated with a cluster) is known to be 60 pc, how far away is B?

Solution:

The magnitude difference between Alpha and Beta is (8 - 3) = 5 magnitudes, which corresponds to a brightness ratio of 100. (E.g. $(2.515)^5 = 100$) so Alpha is apparently 100x brighter than B. According to the inverse square law of light, the brightness of a light source diminishes as the square of the distance.

Accordingly:

$[d(Beta)/d(Alpha)]^2 = 100 = B(Alpha)/ B(Beta)$

and

$[d(Beta)/d(Alpha)] = [100]^{1/2} = 10$

and: d(Beta) = 10 {d(Alpha)}

So, Beta must be ten times more distant than Cepheid Alpha.

4) The period can be obtained as the time between two exactly similar phases, or 'peak to peak' on the light curve curve. This is the time from day 7, to day 18 or (18 d - 7 d) = 11 d.

Take the log of the period: log (11) = 1.04

and now use the P-L graph to obtain the corresponding absolute visual magnitude:

We obtain M_v = (-3.8) approx.

V: Problems on Sidereal Time

In this chapter, we return to basic astronomy in simple time-keeping problems using very basic estimation and approximation procedures. For much of the layout we use a timing diagram, such as that in Fig. 1 which shows the observer ('Obs') at longitude L with respect to Greenwich (G, or the Greenwich meridian at 0 degrees). The L.S.T. or local sidereal time is 6h 00m as denoted by the fact that 6h Right Ascension (RA) is on the observer's meridian. Then the hour angle (HA) for the star will be as shown. (Note: the view shown is a POLAR one, i.e. the Earth as seen from above Earth's north pole).

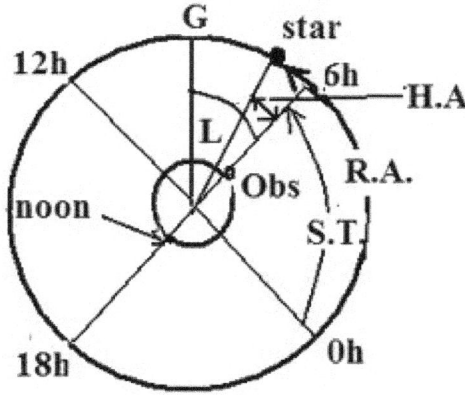

Fig. 1: Time relationships for Observer at longitude L at 6h 00m L.S.T. and with a star at an hour angle H.A. as shown with R.A.

The Right Ascension (RA) is measured in units (hours, minutes, seconds) associated with time, and the Right Ascension of the star is clearly equal to the local sidereal time (L.S.T.) plus the hour angle (which in this case is negative). Thus, we can write:

HA = (RA of observer meridian) − (RA of object)

Again, the hour angle is the angular distance in hours separating your local meridian and the RA of the object. It increases westwards, unlike RA - which increases eastwards. As may be seen from the preceding, the Right Ascension of the observer's meridian (6 h in Figure 1) is none other than the local sidereal time. If you know the RA on your meridian you certainly will know the local sidereal time or L.S.T. Another useful form for many time problems is:

Local time of transit (L.T.T.) = Star's RA - Sun's RA

where L.T.T. is the local mean solar time of transit (or L.M.T.) of the star or other object and Sun's RA is the Sun's Right Ascension on the particular date. A simple interpolation method is adequate (for most purposes) to obtain the latter, given the base data for the specific dates of the solstices and equinoxes below:

Date: March 21-----June 22----Sept. 23-------Dec. 22

Sun R.A.: (0 hr.)-----(6 hr.)------(12 hr.)------(18 hr.)

March 21 is just the (approx.) date of the Vernal Equinox, June 22 is the summer solstice, Sept. 23 the autumnal equinox and Dec. 22 the winter solstice. Since the Earth complete one circuit of the Sun in 365 days, the Sun (as seen from the Earth) will be apparently displaced along the ecliptic (the plane of the Earth's orbit projected onto the celestial sphere) by about one degree per day or equivalent to a 4 minute time difference. (Since 24 hours of Right

Ascension corresponds to a 360° angular difference).

Thus, using either of the four known RA dates for the Sun and knowing the Sun's RA changes by 4 mins./day, the new solar RA on any day can be found.

Example(1):

Find the Right Ascension of the Sun on July 1st.

Solution: We note that 8 days separates July 1st from June 22 when the Sun's R.A. is known to be 6h 00m.

The time difference: 8 days x (4 min/day) = 32 mins.

So Sun's RA on July 1st = 6h 00m + 32 min = 6h 32 m

Example(2):

Aldebaran (RA = 4h 34m) is found to be 45 degrees east of your meridian. What is your L.S.T.?

Solution

First, find the hour angle (HA) of Aldebaran. We have 45° = 45° /(15° h^{-1}) = 3 h. But since the HA is measured *east* of the meridian then one must use the negative value (-3 h) so:

24 h + (- 3h) = 21 h 00 m

We know:

HA = RA of the meridian - RA (object)

and RA of the meridian = L.S.T.

Therefore:

L.S.T.= HA + RA of object

L.S.T. = 21h 00m + 4 h 34 m = 25h 34m = 1h 34m
(25h 34m - 24h 00m)

Other problems:

(1) The Right Ascension of Dubhe in Ursa Major is 11h 02m. Its declination is 61 deg 56 minutes. From a ship at sea it's observed to transit 42 degrees 30 minutes *north of the zenith* on April 8, at 1 a.m. according to the ship's chronometer (which reads Greenwich Mean time or G.M.T.)

Find the position of the ship.

(2)(a) Using a clear diagram, show that the hour angle of a star is related to the Star's RA, the Sun's RA and L.M.T. (local mean time) by:

RA (star) = Sun's RA - (HA) + L.M.T.

(b)Shaula (RA = 17h 31m) transits at 10 p.m. on June 29th. Find the hour angle it exhibits from your location (local hour angle).

(3) The star Canopus (RA = 6h 20m) is observed to have a local hour angle = 45 deg on Feb. 10th for a given location.

(a) What is the local sidereal time.

(b) At what local mean time and standard time would Canopus transit?

(c) What is the approximate L.S.T. at noon on the same date?

(4) Study the diagram for Fig. 1. Using the diagram and inferences regarding time - including Sun's changes in Right Ascension, give the date for which the diagram is referred. If the star in Fig. 1 has an RA = 7h 10m then find its hour angle for the observer (Obs). Hence, or otherwise, obtain the longitude of the observer.

(5)(a) Regulus (10h 07m) is observed to transit on a given date. Given the same date, what would be the local hour angle for: (i) Denebola (RA = 11h 47m), and (ii) Arcturus (RA = 14h 14m)?

(b) Find the L.S.T. one half way between the transits of Denebola and Arcturus. How much later would these two stars set than Regulus?

(c) What RA circle must be rising when Arcturus is setting?

Solutions:

1) If Dubhe transits 42 deg 30 min north of the zenith then its meridian zenith distance (MZD) is 42° 30 m

The ship's latitude may be found from:

Lat. = δ - MZD = 61° 56 m - 42° 30 m = 19° 26 m, North

To obtain the longitude we require:

Greenwich time of transit - local time of transit, or GTT - LTT for short.

Where: LTT = Star's RA - Sun's RA

Now, on April 8, the Sun's RA = 18 days x (4 min/day) = 72 mins. = 1 h 12 min

(Note: this is 18 days elapsed after March 21 when the Sun's RA = 0h)

Then:

LTT = 11 h 02 min - 1 h 12 min = 9h 50 min

According to the ship's chronometer, the star transits at 1 a.m. (13 h) therefore GTT = 13h 00m

Then the ship's longitude is:

Long. = GTT - LTT = 13 h 00 min - 9h 50 min = 3 h 10 min

Converting this to degrees (1 h = 15 deg), we get 47½ degrees. This must be WEST longitude since the GTT *is later than LTT.*

Then the coordinates are:

Lat. = 19 ° 26 m N, Long. = 47½ ° W

(2)(a) Using a clear diagram (below) we show that the hour angle of the star is related to the Star's RA, the Sun's RA and L.M.T. (local mean time) by:

RA (star) = Sun's RA - (HA) + L.M.T.

Solution

This is shown in Fig. 1:

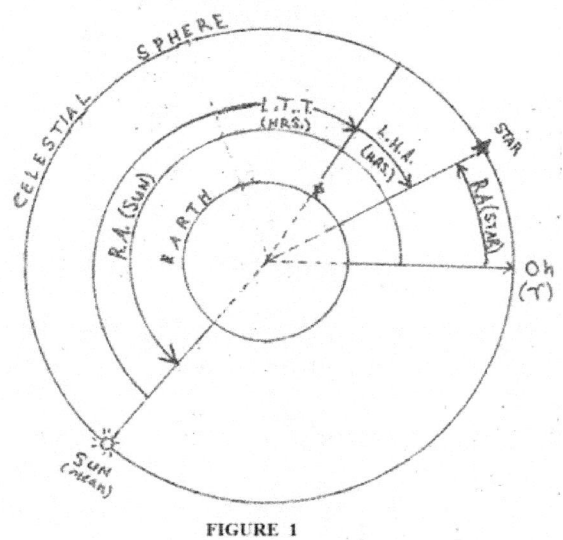

FIGURE 1

(b) Shaula (RA = 17h 31m) transits at 10 p.m. on June 29th. Find the hour angle it exhibits from your location (local hour angle).

Solution:

This can be done from either the diagram or the expression (RA (star) = Sun's RA - (HA) + LMT) which really amount to the same thing.

A simpler tack is to use: LST = HA + RA of object

and it can be ascertained the LST for the date is 15h 31 m

SO: HA = LST - RA = 15h 31m - 17h 31m = -2h 00m, or - 30 deg

(3) The star Canopus (RA = 6h 20m) is observed to have a local hour angle = 45 deg on Feb. 10th for a given location.

(a) What is the local sidereal time?

(b) At what local mean time and standard time would Canopus transit?

(c) What is the approximate LST at noon on the same date?

Solutions to (3):

(a) Again: HA = L.S.T. - RA

So: LST = HA + RA = 3h 00m + 6h 20m = 9h 20m

(b) We have:

LTT = Star's RA - Sun's RA

Feb. 10th is 39 days before March 21 (assume non-leap year) so:

Sun's RA = 0h - 39 days x (4 min/day) = 0h - 156 mins = 0 h - 2h 36 m

Sun's RA = 21h 24m

LTT = 6h 20m - 21h 24m = -17h 36m

Or: 24 h 00m - 17h 36m = 6h 24m

(c) From (a) the L.S.T. on the date is 9h 20m, and this is 9h 20m past noon. Since noon is 9h 20m earlier, then L.S.T.(noon) = 6h 20m - 9h 20m = -3h 00m, or:

24h 00m - 3h 00m = 21h 00m.

(4) Study the diagram for Fig. 1. Using the diagram and inferences regarding time - including Sun's changes in Right Ascension, give the date for which the diagram is referred. If the star in Fig. 1 has an RA = 7h 10m then find its hour angle for the observer (Obs). Hence, or otherwise, obtain the longitude of the observer.

Solution:

The diagram is shown again in Fig. 2 below.

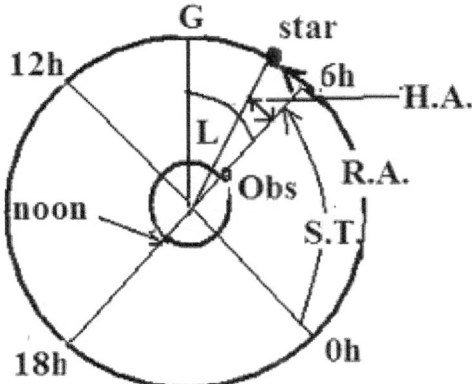

Fig. 2: Time relationships for Observer at longitude L at 6h 00m L.S.T. and with a star at an hour angle H.A. as shown with R.A.

Since the local sidereal time (LST) at the observer location is 6h then this is the RA on the meridian. But the Sun is on the meridian at the antipode or at 18h 00m. Hence, the date must be Dec. 23rd or the winter solstice, since the same RA coincides with the Sun.

If the star's RA = 7h 10m we can find the HA from the angular relationship as:

HA = RA − L.S.T. = 7h 10m - 6h 00m = 1h 10m = 17 .5 deg

The longitude of the observer is: L = G.S.T. − L.S.T. = 9h 00m - 6h 00m = 3h 00m

Converting to degrees: L = 45 deg (W) (since G.S.T. > L.S.T.)

(5)(a) Regulus (10h 07m) is observed to transit on

a given date. Given the same date, what would be the local hour angle for: (i) Denebola (RA = 11h 47m), and (ii) Arcturus (RA = 14h 14m)?

(b) Find the LST one half way between the transits of Denebola and Arcturus. How much later would these two stars set than Regulus?

(c) What RA circle must be rising when Arcturus is setting?

Solutions to (5):

(a) We use: LHA (local hour angle) = RA on meridian (LST) - RA(star)

Here, RA meridian = RA (Regulus) = 10h 07m

Therefore:

LHA (Denebola) = 10h 07m - 11h 47m = -1h 40m (or 22h 20m)

LHA (Arcturus) = 10h 07m - 14h 14m = -4h 07m or 19h 53m

(b) RA(Arcturus) - RA(Denebola) = 14h 14m - 11h 47m = 2h 27m

Then the LST midway between their transits is found by interpolating:

11h 47m + ½(2h 27m) = 13h 00.5m

Thus, Denebola would set about 1h 40m after

Regulus. (E.g. 10h 07m + 1h 40m = 11h47m)

(c) When Arcturus is setting, the RA circle rising is: 14h 14m + 12h 00m = 26h 14m

Or: 26h 14m - 24h 00m = 2h 14m

Additional Exercises:

1) The star Acrux (Alpha Crucis) at RA = 12h 21m, is observed to have a local hour angle = 30 deg on May 10th for a given location.

(a) What is *the local sidereal time.*

(b) At what local mean time and standard time would Acrux transit?

(c) What is the approximate LST at noon on the same date?

2) Between 6 p.m. and 7 p.m local time on March 21st, a star is observed to move from A to B (see diagram below) . H is the observer's horizon. The Sun set directly opposite point A twelve hours earlier (seen from Longitude 90° W)

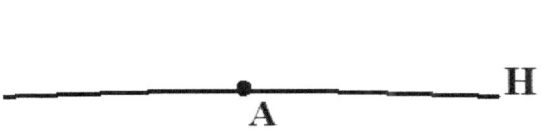

a) What would be the predicted altitude of the star when it gets to B?
b) Estimate the star's azimuth and direction. What is the direction of point B with respect to point A?
c) Where is the vernal equinox in relation to point B? To point A?
d) What is the Right Ascension of the star in the diagram?
e) At approximately what sidereal time would the star reach the meridian?
f) At what local mean time would it reach the meridian?
g) At what Greenwich Mean Time does it reach the meridian?
h) State the cause of the motion from A to B.

3) You are required to calculate the hour angle of the Sun (HA ☉) from a place with longitude 163 ° 14' E. The observation is made at a standard time of 8:46 a.m. on March 10, the standard time being referred to the meridian of longitude 165 ° E. (The Equation of Time for the particular date is approximately E = - 10 min. Note that Equation of Time is just the difference (Apparent time – Mean time).

VI: Assorted Keplerian Orbit Problems

In this chapter we will examine a number of short answer questions related to Keplerian orbits, i.e. those for which one or other of Kepler's three planetary laws had to be applied.

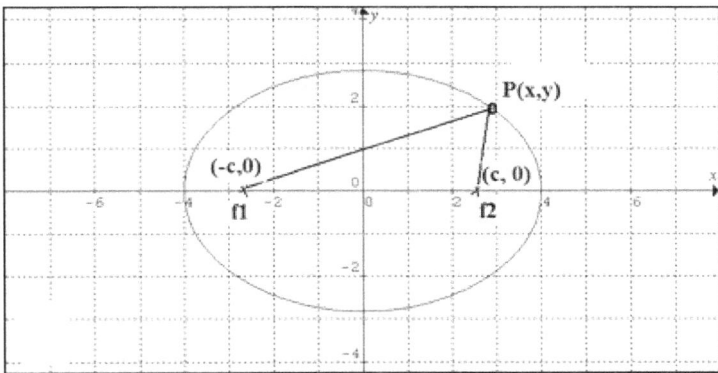

As an example consider the diagram above which shows an ellipse said to be representative of an asteroid orbit. The sketch shows a point P(x,y) on the orbit and also the two foci, f1 and f2. An example problem is to find the eccentricity (e) of the orbit, given only the information evident in the graph.

How to proceed?

First one must be aware of the assorted parameters for the ellipse. There are basically three: a (the semi-major axis), b (the semi-minor axis) and c, the distance between the foci or c = f1 f2.

What is c, mathematically?

It's the square root of the difference between the square of the semi-major axis and semi-minor axis:

$c = (a^2 - b^2)^{1/2}$

Note that when c = 0 we have a = b. (One constant fixed radius: e.g. r = a = b) We can also demonstrate these properties using a simple construction, e.g.

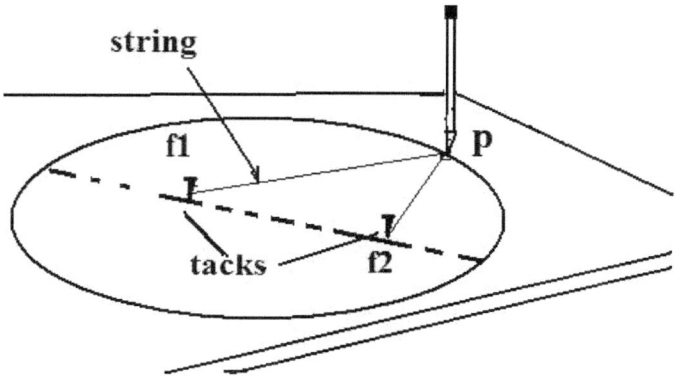

This means if a = b the figure is a perfect circle. When a > b then the shape alters to elliptical.

Again, in general, since the shape of an ellipse requires TWO descriptors, or parameters- one needs to know BOTH a and c to calculate the eccentricity, e, viz.:

$e = c/a = (a^2 - b^2)^{1/2} / a$

Solution:

Now, from the graph, one obtains a = 4 cm and b = 2.8 cm, so c = 2.85 cm. Then to obtain e:

$e = (4^2 - (2.8)^2)^{1/2} / 4 = 0.71$, which is highly elliptical as an asteroid orbit would be.

Then, there is Kepler's Third Law of Planetary Motion or *'The Harmonic law,'* can be written out in terms of:

$(P_1/ P_2)^2 = k(a_1/ a_2)^3$

That is, the squares of the periods (P_1, P_2) of revolution of two planets (being compared) equal k times the cubes of their semi-major axes (a_1, a_2). The last is also known as the 'mean distance from the Sun'. Usually - to make things simpler- we adopt Earth values for P_2 and a_2, and convert the other planet's (that we wish to find) for P_1 and a_1. In the case of the (other) planet's distance, we therefore want to solve for a_1.

Here is a standard template by which to approach the problem:

We take P_2 = 1 year (the period for Earth to make one revolution). We take a_2 = 1 AU or astronomical unit (= 93 million miles). Then the equation simplifies to:

$(P_1/ 1 \text{ yr})^2 = k(a_1/ 1 \text{ AU})^3$

with k a constant of proportionality that can be set equal to 1. Or, on solving for a_1 (and remembering the units used):

$a_1 = \{[P_1]^2\}^{1/3}$

Example Problem:

If for Mars, P_1 = 687 days, find its mean distance from

the Sun (note mean distance taken to be equal to the semi-major axis, a).

Solution:

First convert Mars' period P1 to years, given one Earth year = 365 days. Therefore: P1 = (687 d/ 365 d/yr) = 1.88 years.

Then, solving for a1:

a1 = $\{[1.88]^2\}^{1/3}$ = $\{3.53\}^{1/3}$ = 1.52

I left out units, but we understand that a1 is in AU so

a1 = 1.52 AU

This can also be converted into kilometers:

a1 (km) = 1.52 A.U. x (1.49 x 10^8 km/AU) =

2.26 x 10^8 km

Some Short Answer Problems:

1) A CXC student draws a sketch (Fig. 1) of what he asserts is an asteroid orbit. Using the graph shown, find:

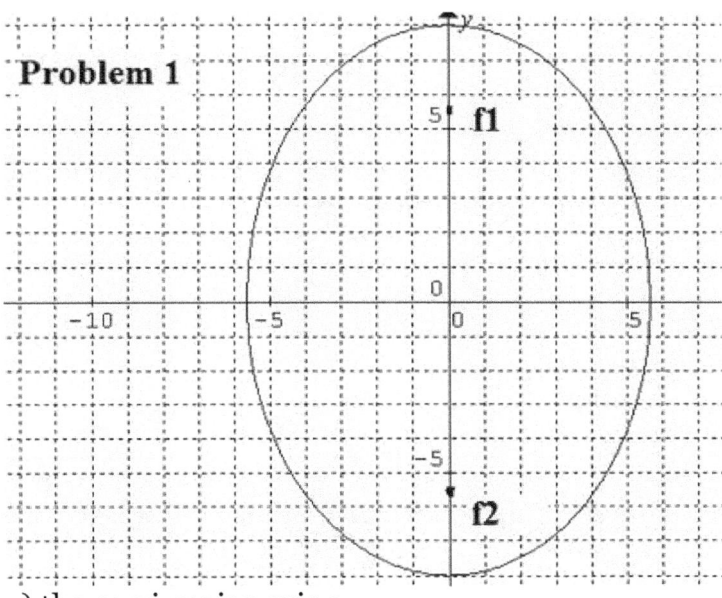

Problem 1

a) the semi-major axis a

b) the semi-minor axis b

c) the eccentricity e.

2) Given the same student orbit sketch, if we let the dimensions be astronomical units (AU) instead of centimeters, then use the graph to estimate the period of the asteroid.

3) A solar system 100 light years from Earth has two planets, X and Y. Planet X has a period of 8 Earth years, and Planet Y has a distance from its Sun of 2.8 astronomical units.

a) Using Kepler's 3rd law of planetary motion, find the distance of Planet X in astronomical units and kilometers.

b) Using the same law, find the PERIOD of Planet Y in Earth years.

4) Consider the table shown below, applied to *circular satellite orbits* around the Earth:

Distance from Center of Earth	2 r	3 r	3r/ 2	4 r	5 r/2	5 r
Acceleration of gravity	g/4					

Complete the Table above given that $r = 6.4 \times 10^6$ m. Hence or otherwise, deduce the acceleration of gravity g at a distance 10r.

5) Newton's version of Kepler's 3rd law is usually written as:

$$(m_1 + m_2)P^2 = 4\pi^2/G \, (r_1 + r_2)^2$$

where G is the Newtonian gravitational constant: ($G = 6.7 \times 10^{-11}$ N-m^2/kg^2) and $(r_1 + r_2)$ is the distance between the centers of the two bodies.

Use this to find the period P of the Moon if its mass $m_1 = m_2/80$ where Earth's mass $m_2 = 6.7 \times 10^{24}$ kg and $(r_1 + r_2) = 384,000$ km. Compare this to the value obtained by using Kepler's *simpler version* of the 3rd law with $(r_1 + r_2) = a_1 = 0.0025$ AU.

Selected Solutions to Problems:

(1) Based on the graph shown with scale in cm, we have:

a = 8 cm, b = 5.7 cm

The eccentricity $e = c/a = (a^2 - b^2)^{1/2} / a$

So: $e = [(8)^2 - (5.7)^2]^{1/2} / 8$

$e = [(64 - 32.5)]^{1/2} / 8 = 5.6$ cm/ 8 cm = 0.7

This could also have been found directly from the graph (measuring the distance of f1 or f2 from 0) and thereby obtaining: c = 5.6 cm

So: e = c/ a = 5.6 cm / 8 cm = 0.7

2) Given the same student orbit sketch, if we let the dimensions be astronomical units (AU) instead of centimeters, then use the graph to estimate the period of the asteroid.

Kepler's 3rd law is required, viz.

$(P/P')^2 = k(a/a')^3$

We take P' = 1 year (the period for Earth to make one revolution). We take a' = 1 AU or astronomical unit (= 93 million miles, or Earth's semi-major axis). Then the equation simplifies to:

$(P/ 1 \text{ yr})^2 = k(a/ 1 \text{ AU})^3$

with k a constant of proportionality that can be set

equal to 1, provided a is in AU and P in yrs. Or, on solving for a (and remembering the units used):

$a^3 = P^2$

But we know a = 8 AU (subst. AU for cm in the graph). Solving for P:

$P = [a^3]^{1/2} = \{[8]^3\}^{1/2} = (512)^{1/2} = 22.6$ yrs.

3) The key to the solution is to insert 'Earth" into the solar system and thereby normalize Kepler's law so the constant of proportionality, k = 1.
E.g.

$(P/ 1 \text{ yr})^2 = k(a/ 1 \text{ AU})^3$

Thus, with appropriate substitutions, we can let P be the period for Planet Y and let a be the distance for Planet X.

To solve for part (a) then, we have: P(X) = 8 yrs, we need a(X):

$a(X) = \{[P(X)]^2\}^{1/3} = [\{8 \text{ yr}\}^2]^{1/3}$

$a(X) = [64]^{1/3} = 4$ AU

(b) Since: $(P(Y) / 1 \text{ yr})^2 = k(a(Y)/ 1 \text{ AU})^3$

where a(Y) = 2.8 AU

Hence:

$P(Y) = [a(Y)^3]^{1/2} = \{[2.8]^3\}^{1/2} = 4.7$ yrs.

4) The solution here is based on noting that the gravitational field intensity $g \sim 1/r^2$

That is, it is *inversely proportional to the distance from the center squared*. Hence, the remaining entries in the table as follows, in ascending order:

$g/9$, $4g/9$, $g/16$, $4g/25$ and $g/25$

Given this relationship ($g \sim 1/r^2$) then at a distance $10r$ we have: $g/100$,

Thus, if $r = 6.4 \times 10^6$ m, then at a distance:

$10\, r = 10\, (6.4 \times 10^6$ m$) = 64 \times 10^6$ m $= 6.4 \times 10^7$ m

And given, $g = 9.8$ ms^{-2}, we would find a new gravitational intensity (corresponding to the much greater distance from the center) of:

$g/100 = (9.8$ ms$^{-2})/ 100 = 0.098$ ms^{-2}

In the same way, the other g-values can be computed for the other radii.

VII: Sidereal & Synodic Periods: Mean Motions:

Among the more important considerations for serious amateur astronomers are the sidereal and synodic periods of planets, and translating between them. In general, we define the *synodic period* to mean the time interval for the revolution of a planet as determined from the same phases, which means in turn the same geometrical configurations. (E.g. or between the same configurations, say between the same oppositions, or between "inferior conjunctions" - see Fig.1).

FIG. 1

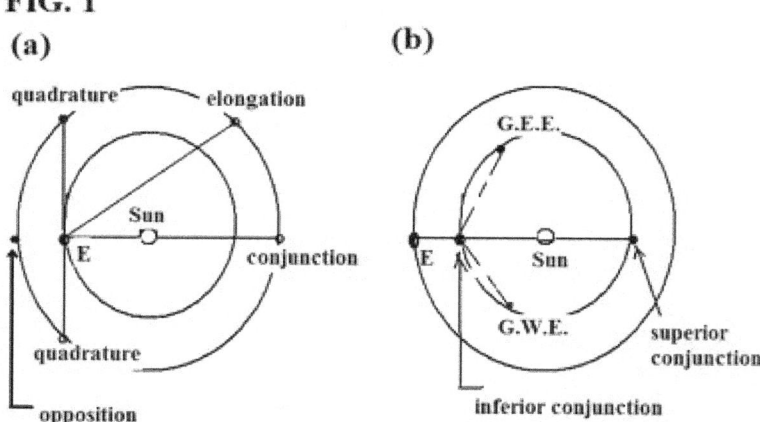

Fig. 1: Configurations for the planets. Right (a) is for superior planets, and (b) is for inferior. G.E.E. refers to greatest eastern elongation, and G.W.E. to western.

As indicated by a study of Fig. 1(a), a superior planet (say Mars) may appear to be 90 degrees away from the Sun in the sky. Therefore, a line from Earth to the Sun makes a right angle with the line extending from Earth (E) to the planet. A planet in this configuration is said to be in *quadrature*. (The planet rises or sets at noon or midnight).

If the superior planet's on the other side of the Sun from Earth, it's then in the same direction from Earth as the Sun and is said to be in *conjunction*. (Note: for the inferior planet, e.g. Venus in Fig. 1(b) the same direction implies *inferior conjunction*).

Another aspect or configuration is referred to as *elongation*. This is just the angle formed between the Earth-planet direction and the Earth-Sun direction. In other words, this elongation is the angular distance from the Sun as seen from the Earth. In the case of the inferior planets (such as Venus and Mercury) one will have a maximum western elongation, and maximum eastern elongation, as shown.

Note also that a superior planet at conjunction has an elongation of 0 degrees, and ditto for the inferior planet at inferior conjunction. Meanwhile, the superior planet has an elongation of 180 degrees at opposition, and 90 degrees at quadrature. An inferior planet, as can be discerned from the geometry in 1(b), can never ever be at opposition. It can, however, be at *superior conjunction*, when it is on the opposite side of the Sun from Earth.

The *sidereal period* is the time interval referencing a planet's revolution about the Sun, reckoned against the fixed stars. We photograph a planet at point 'x' in its orbit, and also with the stars determining a specific background format, then again when it reaches the same point x, relative to that background.

Of particular interest also in these computations are what we call the 'mean motions'. Let P1 and P2 be the sidereal periods of revolution of two planets around the Sun, then the mean motions are defined:

n1 = 360 deg/ P1

n2 = 360 deg/P2

where we assume for example, that n1 > n2

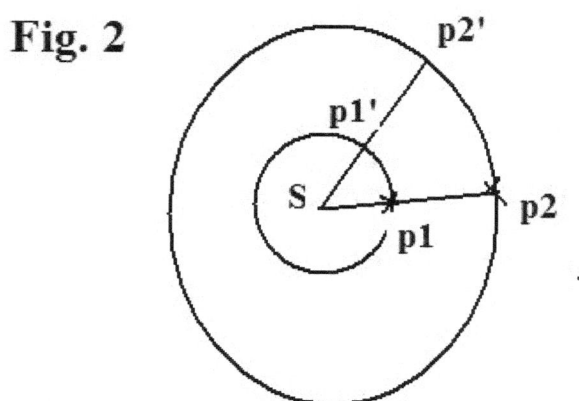

Fig. 2

in which case the radius vector for the (interior) planet p1 gains on the radius vector for planet p2.

As an example, let us consider the earth (P2 = 365¼ days) vs. Venus (P1 = 224.7 days).

Then:

n2 = 360 deg/ (365¼ days) = 0.985647 deg/day

n1 = 360 deg/(224.7 days) = 1.602130 deg/day

which confirms that a planet nearer the Sun has a smaller period of revolution than one further away so that P1 < P2, and also: n1 > n2.(BY VIRTUE OF THE SMALLER PERIOD IT HAS A MORE RAPID MEAN MOTION!)

In effect, in the case above, Venus' radius vector gains on the Earth's by (n2 - n1) degrees per day, or [(1.602130 deg/day) - (0.985647 deg/day)] = 0.61646 deg/day.

By inspection of Fig. 2, let the alignment Sp1p2 denote the positions of the Sun, Venus and Earth at a particular time or "epoch" in the parlance of celestial mechanics. Then, it is clear on the same inspection that one synodic period S will have elapsed by the time the two are in the alignment Sp1'p2'. So, during the time interval S1 (for Venus) its radius vector will have gained 360 degrees on the Earth's.

Now, since Sp1 gains on Sp2 by 0.61646 deg/day then it gains 360 degrees in a time S =

S x (0.61646 deg/day) = 360 deg

Or: S = 360 deg/ (0.61646 deg/day) = 583.9 days

Which is exactly the synodic period for Venus!

Hence, in general:

S x (n1 - n2) = 360 deg or

S = 360 deg/ (n1 - n2)

Or, by reference to the earlier formulations for the n1, n2 mean motions:

$S(360/P_1 - 360/P_2) = 360$ deg

And:

$S = 360$ deg$/[360/P_1 - 360/P_2]$

Or:

$1/S = 1/P_1 - 1/P_2$

In most problem-solving sets the planet's period is sought when Earth's is known. The key to the solution lies in determining whether the unknown planet is an inferior one (e.g. interior to Earth's orbit) or is a superior one, e.g. exterior to the Earth's orbit. Thus there emerge two cases:

(a) The planet is inferior (like Venus), then P_1 refers to the planet's sidereal period and P_2 refers to Earth's.

(b) The planet is superior (e.g. Jupiter) then P_1 refers to the Earth's sidereal period and P_2 to the planet's.

Example Problem:

The synodic period of Venus is found to be 583.9 days. If the length of the Earth's year is 365¼ days, find the sidereal period of Venus:

$1/S = 1/P_1 - 1/P_2$

Venus is an inferior planet, so that Earth's period is inserted for P2.

Then:

1/(583.8 d) = 1/P1 - 1/(365¼ d)

Or:

1/P1 = [365¼ + 583.9 d]/ [583.9 x 365¼]

yielding P1 = 224.7 days

Other problems:

(1) What would be the sidereal period of an inferior planet that appeared at greatest western elongation exactly once a year?

(2) If the synodic period of Saturn is 1.03513 years, find its sidereal period.

(3) A planet's elongation is measured as 125 degrees. Is it an inferior or superior planet?

(4) If the sidereal period of Mercury is 88 days what is its synodic period?

(5) Calculate the ratio of the Earth's mean motion, n, to that of the planet Neptune, given the distance of Neptune is 30.06 AU. (Hint: You will need to apply Kepler's 3rd or harmonic law, see 'Tackling simple Astronomy problems (6)' to first get Neptune's period!)

(6) A newly discovered planet in the Zeta Reticuli system is found to have a semi-major axis of 10 AU. Compare its mean motion n(z) to that of Saturn (hint: reference solution to Problem #2)

Solutions:

1) Recall we define the synodic period to mean the time interval for the revolution of a planet as determined from the same phases, or the same geometrical configurations, i.e. "western elongations". Then the **sidereal period** of an inferior planet that appeared at greatest western elongation exactly once a year would be from:

$1/S = 1/P_1 - 1/P_2$

   Bear in mind if the planet is inferior then P_1 refers to the planet's *sidereal period* and P_2 refers to Earth's.

So:

$1/P_1 = 1/S + 1/P_2$

where: S = 1 yr., P_2 = 1 yr.

then:

$1/P_1 = 1/1 + 1/1 = 2$

so: P_1 = 1/2 yr.

2) Saturn is a superior planet so we apply:

$1/S = 1/P_1 - 1/P_2$

but now with P_1 = Earth's sidereal period, and P_2 = Saturn's. Thus we have:

$1/P_2 = 1/P_1 - 1/S$

where S = 1.03513 years

$1/P_2 = 1/1 - 1/1.03513 = 1 - 0.966 = 0.0339$

$P_2 = 1/0.0339 = 29.4$ yrs.

3) From the diagrams in Fig. 1(a)-(b) of previous blog it is evident that an inferior planet's maximum elongation is still less than 90 degrees - i.e. when its geocentric radius vector is tangential to its orbit configuration. By contrast, the elongation of a superior planet can vary from 0 deg at conjunction to 180 degrees at opposition.

Hence the planet must be a superior one.

4) Again, we employ:

$1/S = 1/P_1 - 1/P_2$

Bear in mind, that since Mercury is inferior then P_1 refers to the planet's sidereal period and P_2 refers to Earth's.

Then: P_1 = 88 d and P_2 = 365¼ d

So:

$1/S = 1/88 - 1/(365¼ \text{ days})$

$1/S = 0.01136 - 0.00273 = 0.00863$

$S = 1/(0.00863) = 115.8 \text{ days}$

5) We are comparing the mean motions, viz.

$n_1 = 360°/P_1$

$n_2 = 360°/P_2$

and let $P_1 = 365¼$ days = 1 yr. for Earth (hence $n_1 = 0.°985647$/day) and P_2 for Neptune is to be found. But we know: a = 30.06 AU

By use of Kepler's 3rd or harmonic law:

$(P_2/P_1)^2 = k(a_2/a_1)^3$

where we let P_1, a_1 have Earth values (a_1 = 1 AU, P_1 = 1 yr.) then if a_2 = 30.06 AU:

$(P_1/1 \text{ yr})^2 = k(30.06 \text{ AU}/1 \text{ AU})^3$

$P = [(30.06)^3]^{½} = 164.8$ yrs.

Then Neptune's mean motion is:

$n_2 = 360°/P_2 = 360°/(164.8 \text{ yrs.}) = 2.°18$ / year

or, in terms of *degrees per day*:

$n_2 = 360 \text{ deg}/ (60193 \text{ d}) = 0°.006 /\text{day}$ (approx.)

Comparing the two by ratio:

$n_1/ n_2 = (0.°985647 /\text{day})/ 0.°006 /\text{day} = 164.8$ x greater

Which could also have been arrived at by taking the ratio (P_2/ P_1).

6) We know already that Saturn's sidereal period = 29.4 yrs. (solution to #2)

For the planet in Zeta Reticuli we must use the same Kepler 3rd law approach used in #5, and we know $a_2 = 10$ AU - let a_1, P_1 be Earth values) so:

$(P(z)/ 1 \text{ yr})^2 = k(30.06 \text{ AU}/ 1 \text{ AU})^3$

$P(z) = [(10)^3]^{1/2} = 31.62$ yrs.

so the planet's mean motion is:

$n(z) = 360° / P(z) = 360° / 31.6 \text{ yr} = 11.°4 /\text{yr}$.

And for Saturn:

$n(S) = 360 \text{ deg}/P(S) = 360 \text{ deg}/ 29.4 \text{ yrs.} = 12.2 \text{ deg/yr}$.

VIII: Relative Distances to Inferior and Superior Planets

One of the more important applications in intermediate astronomy is obtaining the relative distance to an inferior or superior planet in relation to the Earth. As we've seen before, it's easy to apply Kepler's 3rd law to obtain the basic dimensions of an orbit, namely the semi-major axis of the orbit - or the mean distance from the Sun. But things become somewhat more difficult when we seek to find the distance say, of the Earth to the planet, or the planet at a specific time - say before it commences retrograde motion.

A) Inferior Planet's distance   B) Superior Planet's distance

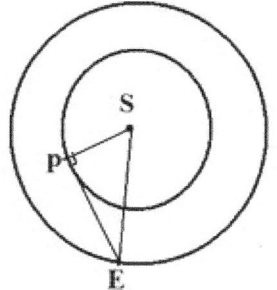   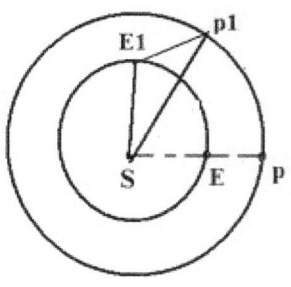

Sp/SE = sin(SEp)   Sp1/ SE1 = sin(p1E1S)/ sin(E1p1S)

In each case above, the distance of the planet can be found in terms of units of Earth's distance from Sun.

So, we consider two cases:

(A) Inferior planet (see the diagram A, left side above)

As seen in the previous chapter, the maximum elongation occurs when the planet's geocentric radius vector (pE in diagram A) is perpendicular to the planet's heliocentric radius vector, pV. Then by a careful measurement of the angle SEp, say over a series of nights around maximum elongation, one can obtain a value for the angle of maximum elongation. At such time the angle SEp is right-angled hence:

$Sp/ SE = \sin(SEp)$

The quantity Sp/SE is the distance of the planet from the Sun in terms of Earth units or AU, (i. e. astronomical units). Let $Sp = R$ and $SE = a(E)$ the semi-major axis for Earth's orbit, then:

$Sp = R = \sin(SEp) [a(E)]$

Example:

If the angle SEp = 60 deg, find the planet's distance from the Sun.

Then: $\sin(SEp) = \sin(60) = \sqrt{3}/2 = 0.866$

So: $R = 0.866[a(E)] = 0.866$ AU

Case (B) Superior Planet

In this case, we apply diagram (B), showing the planet at two successive positions, p and p1 and the Earth at E and E1. This is a more difficult case but can be worked if the planet's synodic period S is known.

Here, we let the planet p be in opposition at some given time with the Earth and Sun (e.g. showing the alignment S-E-p in diagram (B). As we know, with opposition, the elongation is a straight angle or 180 degrees. Then after t days have elapsed the Earth's radius vector SE has moved ahead of the planet's as shown in comparing SE1 to Sp1. As can be seen, this reduces the angle of elongation from 180 degrees at opposition to angle SE1p1. This is then measured.

Now, over t days, the angle ESp will have increased from 0 (at opposition) to a value Θ given by:

$\Theta = [n - n_p] t$

where n, n_p are the mean daily motions of the Earth and the planet, respectively. Using relations for the periods seen in the previous chapter, we may write:

$\Theta = 360 (1/P - 1/P') t$

where P and P' are the sidereal periods for the Earth and the planet, respectively/ Then, it follows by the relations seen already in Chapter VII.

$\Theta = 360 (1/S)$

So, since t and S are both known, Θ can be obtained - that is, angle E1Sp1 is calculated. Hence, angle E1p1S can be found from:

angle E1p1S = 180 - angle SE1p1 - angle E1Sp1

From plane trigonometry we then obtain:

sin(p1E1S)/ Sp1 = sin(E1p1S)/SE1

or:

$Sp1/ SE1 = \sin(p1E1S)/ \sin(E1p1S)$

again, giving the distance from the Sun in terms of Earth's distance unit.

Problems:

1) Estimate the distance of Venus from the Sun at its most recent maximum elongation, if the angle of max. elongation was 46 degrees.

2) A recent observation of Mars 36.5 days after opposition showed an angle of elongation = 136 degrees. Find the distance of Mars from the Earth if Mars' orbital period = 687 days.

Solutions:

PROBLEM 1 Depiction:

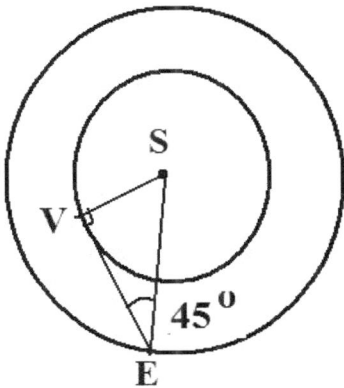

The diagram for Problem 1 is shown on the preceding page, and the angle of maximum elongation.

We know, from the geometry:

SV/ SE = sin (SEV)

Then Venus' distance is:

SV = (sin(SEV)) SE

where: angle SEV = 46 deg and SE = a(E) = 1 AU

then:

SV = sin (46) 1 AU = (0.719) 1 AU = 0.719 AU

Solution for #2

The diagram for the configuration is shown below:

Problem #2 Depiction

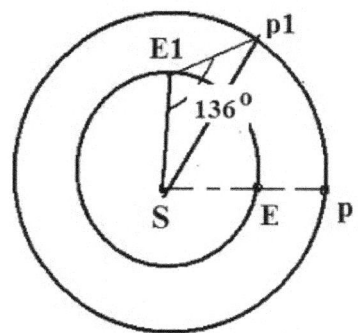

Sp1/ SE1 = sin(p1E1S)/ sin(E1p1S)

We are given t = 36.5 d, and we need to find the "gain" angle:

$\Theta = 360 \, (1/P - 1/P') \, t$

where P, P' are the sidereal periods, or P = 365.25 d (Earth) and P' = 687 d (Mars)

Then:

$\Theta = 360 \, (1/365.25 - 1/687) \, (36.5) = 16.^{\circ}8$

which is none other than angle E1Sp1.

Now, angle **E1p1S** is required in order to get the distance - as the information at the bottom of the depiction shows.

We know:

Ep1S = 180 - [SE1p1 - E1Sp1]

and SE1p1 = 136 °

So:

Ep1S = 180 - [136 ° - 16.° 8]= 27.° 2

Again, from plane trigonometry:

sin(p1E1S)/ Sp1 = sin(E1p1S)/ SE1

and, the ratio of planetary distances at time t:

Sp1/SE1 = sin (p1E1S)/ sin (E1p1S)

But:

$E_1p_1S = 27.°2$

and $p_1E_1S = SE_1p_1 = 136°$

Thus:

$Sp_1/SE_1 = \sin(136)/\sin(27.2)$

$Sp_1 = (0.694/0.457)\, a(E) = 1.518\, (a(E)) = 1.518\text{ AU}$

IX: *Looking at Retrograde Motion*

Recall from Kepler's 3rd or harmonic law:

$(P_1/P_2)^2 = k(a_1/a_2)^3$

where P1, P2 are the periods, related to a1, a2 - the semi-major axes, as shown.

Now, it should be clear that once the sidereal period P of a planet is known, and also the semi-major axis a (or mean heliocentric distance) then the velocity of the planet in its orbit (assumed circular) can be computed, or:

$V = 2\pi a/P$

Hence, for two planets, the ratio of their orbital velocities is:

$V_2/V_1 = (a_2/a_1)(P_1/P_2)$

where we intentionally allow the numbers 1 and 2 to refer to the inner and outer planets, respectively. As may deduced form Kepler's 3rd law:

$P_1 = [(a_1)^3/k]^{1/2}$ and $T_2 = [(a_2)^3/k]^{1/2}$

Substituting for T1 and T2 in the earlier form:

$V_2/V_1 = (a_1/a_2)^{1/2}$

In Fig. 1, the orbits for the Earth and a superior planet are shown, and the semi-major axes are denoted by a and b, respectively. For any superior planet, b > a.

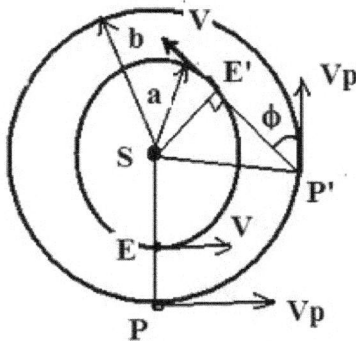

FIG.1: Superior planet & retrograde motion

At opposition (the alignment SEP) the positions of Earth and planet are given as E and P, with velocity vectors V and V_p, tangential to their orbits. Now, from the expression for (V_2/V_1), $V_p < V$ so the angular velocity of the planet as observed from Earth is: ($V_p - V$)/ PE and is in a direction opposite to the orbital motion, and hence is *retrograde* at opposition.

At the following quadrature, shown by configuration SE' P', the Earth's orbital velocity V is now along the line P'E' but the planet's velocity V_p has a component $V_p \sin(\varphi)$ perpendicular to E'P'. The other component, $V_p \cos(\varphi)$ lies along the line P'E' and - like the Earth's velocity - doesn't contribute to the observed angular velocity of the planet.

The geocentric angular velocity at quadrature is then:

$V_p \sin(\varphi)$/ E'P'

Example Problem:

a) Compare the orbital velocities of Venus and Earth, if the sidereal period for Venus, T_1, is 224.69 d, and for Earth (T_2) is 365.25 d.

b) Verify this by using a Table of orbital velocities for the planets - given in km/s

Solution:

We have: $V_2/V_1 = (a_2/a_1)(T_1/T_2)$

By convention we assign '1' to the inner planet (Venus) and '2' to the outer (Earth). We have $a_2 = 1$ AU and for Venus (from Kepler's third law):

$T_1 = (224.69/365.25)$ yr. $= 0.6151$ yr.

$a_1 = \{[T_1]^2\}^{1/3} = [(0.6151)^2]^{1/3}$

$a_1 = 0.723$ AU

Therefore:

$V_2/V_1 = (0.7234)(1/0.6151)$

$V_2/V_1 = 1.175$

(b) According to a Table of Orbital Velocities in ***Astrometric & Geodetic Data***:

V(Venus) = 35.02 km/s

V(Earth) = 29.78 km/s

Take the ratio (which must be 1.175)

V(Venus)/V(earth) = (35.02 km/s)/ (29.78 km/s) = 1.175

So, Venus' orbital velocity is 1.175 times Earth's

Other Problems:

(1)(a) Calculate the ratio of the Earth's tangential orbital velocity to Saturn's given Saturn's sidereal orbital period is 10,746.9 days.

(b) Validate this is approximately correct if Saturn's mean orbital velocity is 9.664 km/s from an astrometric table.

(2) At quadrature with the planet Mars, it is found (based on Fig. 1) that SP' = 1.53 AU, and SE' = 0.999 AU.

(a) From this deduce the distance P'E' and

(b) Hence, find the angular velocity of the planet *as observed from Earth*

(3) If Mercury's sidereal orbital period = 0.2408 yr. show that its tangential orbital velocity should be 47.87 km/s

(4) Based on the information in (3) and given that Uranus' sidereal period = 83.747 yrs., show that

Uranus tangential orbital velocity is approximately 5.477 km/s.

Solutions:

1.(a) We have:

$V_2/V_1 = (a_2/a_1)(T_1/T_2)$

By convention we assign '1' to the inner planet (Earth) and '2' to the outer (Saturn). We have $a_1 = 1$ AU and for Saturn (from Kepler's Third law):

$T_2 = (10{,}746.9/365.25)$ yr. $= 29.4$ yr.

$a_2 = \{[T_2]^2\}^{1/3} = [(29.4)^2]^{1/3} = 9.50$ AU

Whence:

$(V_1/V_2) = (a_2/a_1)^{1/2} = (9.50)^{1/2} = 3.08$

So, the Earth's orbital velocity should be about 3.1 x faster than Saturn's

(b) Saturn's mean orbital velocity is 9.664 km/s from an astrometric table.

From the sample problem, V(Earth) = 29.78 km/s

Then: $(V_1/V_2) = (29.78$ km/s$)/(9.664$ km/s$) = 3.08$

(2) (a) SP' = 1.53 AU is the hypoteneuse of the right triangle, SE'P'. SE' = 0.999 AU is one leg. Then, from Pythagoras' theorem:

$P'E' = [(SP')^2 - (SE')^2]^{1/2} = [1.53^2 - 0.999^2]^{1/2}$

$P'E' = 1.15$ AU

(b) To obtain the angular velocity as observed from Earth we first need to obtain angle SP'E' since we must find φ, knowing that: (90 - S'P'E' = φ). From trigonometry:

$\cos(SP'E') = $ adj/ hyp $= P'E'/SP' = 1.15/1.53 = 0.757$

$SP'E' = \arccos(0.757) = 40.°8$

then: $\varphi = 90° - S'P'E' = 90° - 40.°8 = 49.°2$

The geocentric angular velocity at quadrature is then:

$V_p \sin(\varphi)/ E'P'$

To get V_p, note:

$(V_1/V_p) = (a_p/a_1)^{1/2} = (1.53)^{1/2} = 1.23$

So: $V_p = (1/1.23)(29.78 \text{ km/s}) = 24.2$ km/s

and:

$V_p \sin(\varphi)/E'P' = [(24.2 \text{ km/s})\sin(49.2)]/1.15$
$= 15.9$ km/s

(3) Mercury is interior to Earth so designate with '1':

$T_1 = 0.2408$ yr.

$a_1 = \{[T_1]^2\}^{1/3} = [(0.2408)^2]^{1/3} = 0.387$ AU

$V_2/V_1 = (a_1/a_2)^{1/2}$

So: $V_1 = (V_2)/(a_1/a_2)^{1/2}$

where: V_2 is for Earth (29.78 km/s) and:

$(a_1/a_2)^{1/2} = (0.387/1)^{1/2} = 0.622$

$V_1 \, ☿ = V_2/ 0.622 = (29.78 \text{ km/s})/0.622 = 47.87$ km/s

(4) By convention again we assign subscript '1' to the inner planet (Earth) and '2' to the outer (Uranus). We have $a_1 = 1$ AU and for Uranus (from Kepler's 3rd law): $T_2 = 83.747$ yrs.

By Kepler's 3rd law:

$a_2 = \{[T_2]^2\}^{1/3} = [(83.747)^2]^{1/3} = 19.13$

Whence:

$(V_1/V_2) = (a_2/a_1)^{1/2} = (19.13)^{1/2} = 4.37$

or: V_2 ('Uranus') $= V_1/ (4.37)$

where $V_1 = 29.78$ km/s (Earth) so:

$V_2 = 29.78$ km/s$/ (4.37) = 6.8$ km/s *

Explanation:

A cross check of other references (e.g. Abell, **'Exploration of the Universe'**, Appendix 8) discloses that this is the correct value! The 5.477 km/s from the Astrometric & Geodetic data sheet is incorrect! It actually applies to *Neptune*, not Uranus.

We can check this: The semi-major axis of Neptune is 30.06 AU (from problem #5, of Chapter VI.) By Kepler's 3rd law:

a2 = 30.06 AU

Whence:

$(V_1/V_2) = (a_2/a_1)^{1/2} = (30.06)^{1/2} = 5.48$

or: V_2 ('Neptune') = V_1/ (5.48) = (29.78 km/s)/ 5.48 = 5.43 km/s

It pays to double check!

One of the key tasks for planetary astronomers - amateur or professional - is to ascertain the point at which a superior planet's motion changes from being retrograde to direct. We are referring here to the planet's geocentric (Earth-centered or referenced) angular velocity, and identifying when it becomes direct - between opposition and quadrature (see previous chapters for these configuration definitions).

To fix ideas here, we use the diagram below and let the positions of Earth and planet at the stationary point be denoted by E and P. The velocity of P relative to E must lie along the geocentric radius vector (EP) if the planet appears stationary

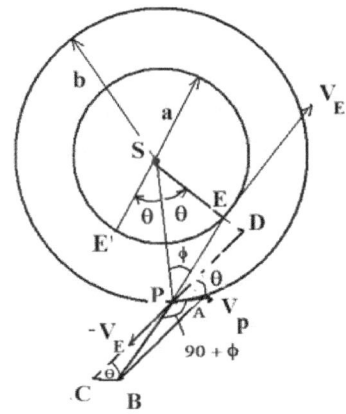

Geometric configuration for a
stationary point- superior planet

Clearly, if such a change occurs, there will be some point where the motion is neither retrograde or direct. This is called a "null" point, or better: "**stationary point**". The problem then, is to obtain an expression for the planet's elongation E at a stationary point in terms of the distances of the planet and Earth from the Sun. Using such an expression and the measured elongation the planet's heliocentric distance can be computed. (In addition, we can also obtain an analytical expression giving the angle Θ between the heliocentric radius vectors of planet and Earth at the stationary point) Further, such an analytic expression together with the synodic period of the planet, allows us to predict its next stationary point.

As we know, we can form a "parallelogram" of velocities - just as we can for forces. In this case, we can form the velocity parallelogram PABC, with the resultant given by PB.(E.g. the resultant of the vectors

($-V_E$) and the planet's tangential velocity V_P. Note that: PA = V_P and PC = $-V_E$.

Since we assume the orbits are circular (a close enough approximation) the angle Θ between the heliocentric radius vectors must be angle PCB between the velocity vectors. In addition, note that angle APB = 90 + φ. Now, if we extend CP to meet the line SE (produced at D) then angle DPA = Θ and angle EPD = 90 - (Θ + φ). Focusing now on the triangle PCB, we may use the sine formula of trigonometry to write:

$\sin[90 - (\Theta + \varphi)] / V_P = \sin(90 + \varphi) / V_E$

or (using trigonometric identities):

$\cos(\Theta + \varphi) = (V_P/V_E) \cos \varphi$

Making use of another trigonometric identity and triangle SEP, we may write:

SP = PE $\cos \varphi$ + SE $\cos \Theta$

or:

(I) b = PE $\cos \varphi$ + a $\cos \Theta$

Applying it once more to triangle SEP we get:

SE = SP $\cos \Theta$ + PE $\cos$ (SEP)

or:

(II) a = b $\cos \Theta$ + PE $\cos(\Theta + \varphi)$

From the geometry of the configuration, since angle SEP = 180 - ($\Theta + \varphi$), we can write equations (I) and (II) as:

(Ia) $b - a \cos \Theta = PE \cos \varphi$

and:

(IIa) $a - b \cos \Theta = PE \cos (\Theta + \varphi)$

Dividing through eqn. (Ia) by eqn. (IIa):

$[b - a \cos \Theta]/ [a - b \cos \Theta] = \cos \varphi / [\cos (\Theta + \varphi)]$

Now, $\cos (\Theta + \varphi) = (V_P/V_E) \cos \varphi$

and from the previous chapter, we saw:

$(V_2/V_1) = (a_1/a_2)^{1/2}$

Combining these in (Ia) and (IIa) we have:

$[b - a \cos \Theta]/ [a - b \cos \Theta] = (V_P/V_E) = (b/a)^{1/2}$

Algebraically re-arranging we find:

$\cos \Theta = [a^{1/2} b^{1/2} (a^{1/2} + b^{1/2})]/ (a^{3/2} + b^{3/2})$

The preceding can be vastly simplified if all units are normalized to those in terms of Earth's to the Sun. Thus, a = 1 and b is in terms of a. So let:

$Q^{1/2} = a^{1/2} b^{1/2}$

$(1 + Q^{1/2}) = a^{3/2} + b^{3/2}$

Then re-write as:

$\cos \Theta = [Q^{1/2} (1 + Q^{1/2})] / (1 + Q^{3/2})$

As may be discerned from the diagram, when Earth is at point E' where angle ESP = 90 degrees, we have another stationary point. It is of interest here to compute the total time during which a planet will be seen to move retrograde and this is just the time it takes Earth's radius vector to advance through an angle equal to 2 Θ, with respect to the planet's radius vector. This will be given by t(Θ), where:

t(Θ) = 2 Θ / 360 x S X (ΘS)/ 180

where S is the synodic period.

The time that elapses between opposition and the next stationary point is just t $_{(\Theta)/2}$.

Meanwhile, the time interval during a synodic period that a planet's motion is direct is t(D) where:

t(D) = (360 - 2 Θ)/ 360 X S X (1 - Θ/ 180)S

Example Problem:

Mars has a synodic period S = 779.9 days. At a particular stationary point Mars' radius vector is determined to be 1.52 AU. Find the angle Θ that Earth would have advanced to reach this point and time elapsed between opposition and the next stationary point.

Solution:

We have:

$$\cos \Theta = [Q^{1/2}(1 + Q^{1/2})] / (1 + Q^{3/2})$$

where:

$$Q^{1/2} = a^{1/2} b^{1/2} = (1)(1.52)^{1/2} = 1.23$$

$$(1 + Q^{3/2}) = a^{1/2} + b^{3/2} = 1 + (1.52)^{3/2} = 1 + 1.87 = 2.87$$

Then:

$$\cos \Theta = [(1.23)^{1/2}(2.23)] / (2.87)$$

$$\cos \Theta = [(1.10)(2.23)] / (2.87) = 1.64/2.87 = 0.571$$

$$\Theta = \arccos[0.571] = 55.^\circ 1$$

The time that elapses between opposition and the next stationary point ($t_{(\Theta)/2}$) is just:

$$t_{(\Theta)/2} = \tfrac{1}{2} [2\Theta / 360 \times S \times (\Theta S)/180]$$

Where: $S = 779.9d / 365.25 d = 2.13$ yrs

So:

$$t_{(\Theta)/2} = \tfrac{1}{2} [(110.2)/360 \times (2.13) \times (55.1 \times 2.13)/180]$$

$$t_{(\Theta)/2} = \tfrac{1}{2} [(0.306)(2.13)(0.652)] = 0.21 \text{ yrs}.$$

Other problems:

1) Mars reaches a stationary point 36 ½ days after opposition. Its elongation is then measured to be 136.2 deg. Given that *the sidereal period* is 687 days, find the distance of Mars from Earth in AU. Also, find the time to its next stationary point.

2) Find the length of time Jupiter has retrograde motion in each synodic period given its heliocentric distance is 5.2 AU and its sidereal period is 11.86 years.

3) In the Epsilon Eridani star system a planet designated Epsilon Eridani III is determined to have the exact same orbital parameters as Earth (e.g. a, e, i etc.).

In the same system, another exo-planet designated Epsilon Eridani IV is found to have V_P = 4.5 km/s.

a) Using your knowledge of the known parameters, plus the diagram shown, construct an appropriate parallelogram of velocities and hence obtain the angles: Θ and φ.

b) Hence, or otherwise, estimate the time planet Epsilon Eridani IV will be moving retrograde relative to Epsilon Eridani III, and also the time between its opposition and the next stationary point.

c) Obtain the time during Epsilon Eridani IV's synodic period that it is moving direct.

Solutions:

1) Use: $1/S = 1/P_1 - 1/P_2$

   First obtain Mars synodic period, S:

   with P_1 = Earth's sidereal period (1 yr), and P_2 = Mars $(687/365.25) = 1.88$ yr

   $1/S = 1 - 1/1.88 = 1 - 0.5319 = 0.468$ yr

   Then: $S = 1/0.468 = 2.13$ yrs. (or 779.9 days)

   Mars' distance from Earth is PE from the diagram, so may be obtained from:

   $b - a \cos \Theta = PE \cos \varphi$

   and: $PE = (b - a \cos \Theta)/ \cos \varphi$

   Now (given $a = 1$ and $b = 1.52$):

   $\cos \Theta = [a^{1/2} b^{1/2} (a^{1/2} + b^{1/2})]/ (a^{3/2} + b^{3/2})$

   $= [(1.52)^{1/2} ((1)^{1/2} + (1.52)^{1/2})]/ ((1)^{3/2} + (1.52)^{3/2})$

   $\cos \Theta = 0.958$

   so: $\Theta = 16.°7$

   And the angle of elongation ($136.°2$) is (in terms of the geometry):

   $136.°2 = 180° - (\Theta + \varphi)$

So we can solve for φ:

φ = 180° − 136.°2 − 16.°7 = 27.°1

Now:

PE = (b − a cos Θ)/ cos φ

= (1.520 − 0.958)/ cos (27.1) = (0.562)/0.890
= 0.631

So the distance of Mars from Earth denoted by PE (at stationary point) = 0.631 AU

The time to next stationary point:

$T_{r/2}$ = Θ/ 360° x S

where: (given a =1, b = 1.52):

Then:

$T_{r/2}$ = Θ/ 360° x S = (16.°7)/360° x (779.9 d)
= 36.1 days

2) First, use: 1/ S = 1/P1 − 1/P2

And first obtain Jupiter's synodic period, S:

Where P1 = Earth's sidereal period (1 yr), and P2 = Jupiter's (11.86 yrs)

1/S = 1 − 1/(11.86) = 1 − 0.0843 = 0.9156

Then: S = 1/ 0.9156 = 1.09 yrs. = 398.1 d

The time for retrograde motion in each synodic period is:

$T_r = 2\Theta/360° \times S$

Where:

$\cos \Theta = [a^{1/2} b^{1/2} (a^{1/2} + b^{1/2})]/(a^{3/2} + b^{3/2})$

and a =1, b = 5.2, so:

$\cos \Theta = [(5.2)^{1/2} ((1)^{1/2} + (5.2)^{1/2})]/((1)^{3/2} + (5.2)^{3/2})$

And: $\cos \Theta = 0.582$

So: $\Theta = \arccos(0.582) = 54.°4$

and the time to move retrograde over each synodic period is:

$T_r = 2\Theta/360° \times S = (2 \times 54.°4)/360° \times (1.09 \text{ yr}) = 0.329$ yr. = 120.3 days

3) We use $V_P = 4.5$ km/s and $V_E = 47$ km/s for the velocities.

Since Epsilon Eridani III has the same orbital parameters as Earth we can employ Earth semi-major axis, etc. in the computations. Also, a check of tables (or previous problems from earlier sets) shows Epsilon Eridani IV has the same orbital velocity as Neptune so will have approximately the same sidereal period of 163.73 yrs. Now to form the parallelogram we need to obtain the angles Θ and φ.

We must first obtain *the synodic period* of Epsilon Eridani IV:

$1/S = 1/P_1 - 1/P_2$

Where $P_1 = 1$ yr (for Epsilon Eridani III), and $P_2 = 163.7$ yrs. for IV

$1/S = 1 - 1/163.7 = 1 - 0.0061 = 0.9939$

Then: $S = 1/0.9939 = 1.006$ yr. $= 367.7$ d

To get Θ: $\cos \Theta = [a^{1/2} b^{1/2} (a^{1/2} + b^{1/2})]/(a^{3/2} + b^{3/2})$

where b is Epsilon Eridani IV's semi-major axis or:

$b = \{[P_2]^2\}^{1/3} = \{[163.7]^2\}^{1/3} = 29.9$ AU

and we know $a = 1$, so:

$\cos \Theta = [(29.9)^{1/2} ((1)^{1/2} + (29.9)^{1/2})]/((1)^{3/2} + (29.9)^{3/2})$

$\cos \Theta = 0.215$ and $\Theta = \arccos(0.215) = 77.°6$

Meanwhile, by using a scaled diagram (e.g. see the example shown below – but best done using large graph paper) we find: $90° + \varphi = 105°$ so:

$\varphi = 105° - 90° = 15°$

Prob. 3 Part (a).
N.B. Diagram Not to scale!

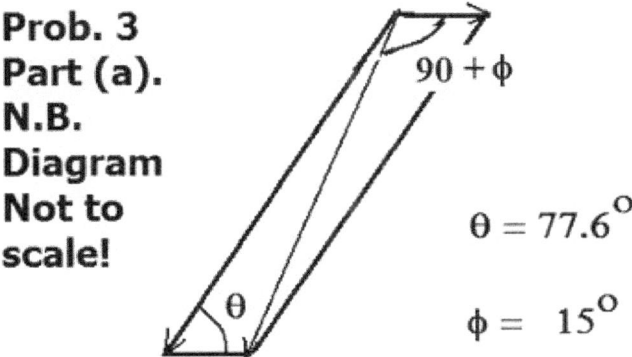

$\theta = 77.6°$

$\phi = 15°$

b) The time for retrograde is:

$T_r = 2\Theta/360° \times S = (2 \times 77.°6)/360° \times (367.7 \text{ d}) = 158.5 \text{ d}$

Between opposition and next stationary point:

$T_{r/2} = (158.5 \text{ d})/2 = 79¼ \text{ d}$

c) The time during Epsilon Eridani IV's synodic period *that it is moving direct* is given by:

$t(D) = (1 - \Theta/180°) \times S = (1 - 77.°6/180°)(367.7 \text{ d}) = 209 \text{ days}$

(N.B. Remember these are times referred to the **synodic** not sidereal periods!)

X: Basic Celestial Mechanics

We're now at the point of tackling true elliptical orbits, as opposed to elliptical approximations. So all the problems set or done will reflect that. We will use the diagram shown below, which denotes a true elliptical orbit defined by some radius vector, r.

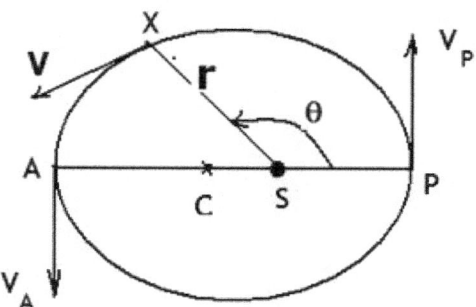

Elliptic orbit basics

This is the variable distance from central object, i.e. Sun, to the orbiting body, i.e. planet, at any given time. It is different from the semi-major axis, a, because the latter is a defined *mean* while the former depends on the actual distance at some time t. Thus, what this Chapter is all about is generalizing the treatment of orbits to be able to work more and varied problems, including those involving planetary satellites and artificial Earth satellites.

Now, let the points A and P denote the aphelion (farthest point) and perihelion (closest point) respectively and V_A, V_P the respective velocities at those orbit extrema. As may be deduced here, points A and P are the only ones in the whole orbit for which the velocities are truly tangential or at right angles to the radius vectors for those positions. Consequently, we can write:

$V = (2\pi/T)\, r$

where r is the radius vector at the point, and T *is the period*.

If Kepler's 2nd law holds at every point (equal areas swept out in equal intervals of time) we also have:

$r^2\, (2\pi/T) = h$

where 'h' is a constant ('specific relative angular momentum') which is twice the rate of area description (i.e. by the radius vector). Thus, if the radius vector is r_1, then $h = 2A_1$, when $A_1 = \pi(r_1)^2$. Hence, at aphelion and perihelion only we have:

$V = h/r$

For the perihelion velocity we have:

$V_P = h/\, a(1 - e)$

where a is the semi-major axis, and e is the eccentricity.

For the velocity at aphelion:

$V_A = h/\, a\,(1 + e)$

Then the ratio of velocities is:

$(V_P/V_A) = (1 + e)/ (1 - e)$

The correct energy equation can be written:

$\frac{1}{2}V^2 - \mu/r = \alpha$

where α is an energy integration constant.

Since for any bound system of masses m1 and m2, $\mu = G(m1 + m2)$, where G is the Newtonian gravitational constant ($G = 6.7 \times 10^{-11}$ Nm²/kg²) then if we know V_P and V_A, along with a and e, we can compute α, viz.

$\alpha = \frac{1}{2}V_P^2 - \mu/a(1 - e)$

at perihelion, and

$\alpha = \frac{1}{2}V_A^2 - \mu/a(1 + e)$

at aphelion.

subtracting the lower equation from the upper and using the ratio of velocities expression (V_P/V_A), we get:

$(V_P)^2 = \mu/a\,[(1 + e)/ (1 - e)]$

and:

$(V_A)^2 = \mu/a\,[(1 - e)/ (1 + e)]$

Combining these, we obtain, much more simply:

$(V_A)(V_P) = \mu/a$

with further manipulation using the energy equation, the ambitious reader should be able to obtain:

$V^2 = \mu (2/r - 1/a)$

which is also known as ***the "vis viva" equation***, one of the most important in celestial mechanics.

Now, what if the planet orbit is fully circular, so r = a?

Then we see the vis viva reduces to:

$V^2 = \mu (2/a - 1/a) = \mu [(2a - a)/a^2] = \mu (a/a^2) = \mu/a$

But as we saw:

$V = 2\pi a/T$

then:

$T = 2\pi (a^3/\mu)^{1/2}$

which holds even when the orbit is elliptical.

Sample Problems:

1) You've discovered a new Jovian moon that takes 30 days to go around Jupiter and is 19 Jupiter diameters from the planet. Deduce Jupiter's mass based on this information?

Solution (One way - there are others!)

The problem basically uses Kepler's harmonic law to obtain Jupiter's mass from Jupiter's moon's motion - knowing the mass of the Sun (1 solar mass). From the harmonic law we know that the period squared is proportional to the semi-major axis of the orbit cubed:

$T^2 \sim a^3$

where T is in years and a in astronomical units or AU. (1 AU = 1.496×10^8 km) Note: when we use AU and yrs. we are implicitly comparing the elements of the unknown system (e.g. moon going round Jupiter) with *the known system of the Earth going round the Sun*. For Earth, we know T = 1 year, a = 1 AU.

For the newly discovered moon going round Jupiter, T(m) = 30 days. Or, in the specific units we need: P(m) = 30 days/ 365 days/year = 0.0822 yrs.

The distance or semi-major axis for the moon's orbit is 19 Jupiter diameters, or a(m) = 19 x (1.43×10^5 km) = 2.717×10^6 km where the quantity in brackets is the known equatorial diameter of Jupiter, found from a table of planetary data.

The value above is a(m) = 0.018 AU since 2.717×10^6 km)/ (1.496×10^8 km/ AU) = 0.018 AU. Now, according to the use of the 3rd law, the relations between period and semi-major axis obtaining for the Earth-Sun ought to be in the same proportion as the relation between period and semi-major axis for the moon of Jupiter and Jupiter. Thus, we can write the proportion: M(S)/ M(J) = 1/ (a^3/ T^2)

Here, M(S) is the mass of the Sun and M(J) is the mass for Jupiter. The numerator on the right side is simply the ratio of a^3 to T^2 for Earth :

$$(1 \text{ AU})^3 / (1 \text{ yr.})^2 = 1 \text{ AU}^3/\text{yr}^2$$

The denominator is for the moon of Jupiter's values, or:

$$(a^3)/T^2 = (0.018 \text{ AU})^3 / (0.0822 \text{ yr})^2 =$$

$$0.00086 \text{ AU}^3/\text{yr}^2$$

Now, put all these values back into the proportion we obtained:

$$M(S)/M(J) = 1/0.00086$$

Note that the units of AU and yrs. cancel out between the numerator (e.g. values for Earth or 1 AU^3/yr^2) and the denominator (values for moon of Jupiter, e.g. $(0.018 \text{ AU})^3/(0.0822 \text{ yr})^2$)

What we have then at the end, is a simple proportion from which we can find the mass of Jupiter (M(J). Cross multiplying and setting the terms equal:

$$M(J) = 0.00086 \text{ M}(S)$$

In other words, the mass of Jupiter is about 8.6×10^{-4} of the Sun's.

Or, since the Sun = 1 solar mass (by definition) then Jupiter must be 0.00086 solar masses.

To check this, consult a table of planetary masses (See Appendix) to find Jupiter's mass = 1.89×10^{27} kg and the Sun's = 1.99×10^{30} kg. Taking the ratio:

$(1.89 \times 10^{27}$ kg$)/ (1.99 \times 10^{30}$ kg$) = 0.0094$, or slightly off.

(2) The Pluto-Charon system, i.e. the orbit of Charon with respect to Pluto, has an eccentricity e = 0.0020. The semi-major axis of the orbit is 19, 450 km. The mass of Pluto = 1.27×10^{22} kg and the mass ratio (Charon to Pluto) is found to be $m(c)/m(P) = 0.12$. From this information, find:

a) The mass of Charon

b) The ratio of the velocity of Charon at perihelion to aphelion

c) The period of Charon, and its velocity

Solution:

a) The mass is found using the mass ratio. Since $m(c)/m(P) = 0.12$ and $m(P) = 1.27 \times 10^{22}$ kg, then:

$m(C) = 0.12 (1.27 \times 10^{22}$ kg$) = 1.52 \times 10^{21}$ kg

b) This ratio is obtained from:

$(V_P/V_A) = (1 + e)/ (1 - e)$

and we know, e = 0.0020, so:

$(V_P/V_A) = (1 + 0.0020)/ (1 - 0.0020) =$

$(1.0020)/ (0.998) = 1.004$

Therefore: $(V_P/V_A) = 1.004$

c) The period is obtained from: $T = 2\pi (a^3/\mu)^{1/2}$

where $a = 19,450$ km $= 1.94 \times 10^7$ m

For preserving unit consistency this is one time we decline to use units of AU and years, and instead use the distance in meters to conform with the required units for G. Then: $T =$

$2\pi [(1.94 \times 10^7 \text{ m})^3/[6.7 \times 10^{-11} \text{ Nm}^2/\text{kg}^2) (1.422 \times 10^{22} \text{ kg})]^{1/2}$

$T = 5.5 \times 10^5$ s $= 6.37$ days

The velocity is:

$V = 2\pi a/T = 2\pi(1.94 \times 10^7 \text{ m})/(5.5 \times 10^5 \text{ s}) =$

221.6 ms^{-1}

Other Problems:

1) Derive the vis viva equation using three or more of the equations given in this blog. (Hint: all incorporate u, and two incorporate e)

2) The Earth's aphelion distance is 1.01671 AU and its

perihelion distance is 0.98329 AU. Given its eccentricity e = 0.016, then use this information and any other (from previous blogs) to find:

a) the velocities at aphelion and perihelion

b) the energy constants C(A) and C(P) at each of these points, and h, the 'specific relative angular momentum'.

c) Hence or otherwise, use the vis viva equation to confirm the results you obtained in (a)

3) Attempt to obtain an improved value for Jupiter's mass (from what the first sample problem yields) using: T = = $2\pi (a^3/\mu)^{1/2}$ (Recall $\mu = G(m_1 + m_2)$ and Jupiter's mass and the Sun's have already been given along with G, in the blog general information)

4) The orbital period of Jupiter's 5th satellite is 0.4982 days about the planet. Its orbital semi-major axis is 0.001207 AU. The orbital period and semi-major axis of Jupiter are 11.86 yrs. and 5.203 AU. Find the ratio of the mass of Jupiter to that of the Sun. Comment on why or why not you expect this result to be different from that in sample problem (1).

5) A communications satellite is in a circular equatorial orbit about Earth and always remains above a point of fixed longitude. If the sidereal day is 23h 56m long and the year 365.25 days in length and the distance of the satellite from Earth's center is 41,800 km, deduce the ratio of the mass of the Sun to the mass of the Earth. Hence, or otherwise, obtain the

mass of Earth in kg if the mass of the Sun is as given in the check of the sample problem (1). (Take 1 AU = 149.5 x 10^6 km)

6) Based on the information you obtained from probs. 4 and 5 (as well as their solutions), compute the change in Jupiter's orbital period if it suddenly became the same mass as Earth.

7) Halley's comet moves in an elliptical orbit of e= 0.9673. Calculate: a) the ratio of its linear velocities, and b) its angular velocities at perihelion and aphelion. Is it possible from this information to obtain *the semi-major axis* for this comet? If yes, then proceed to compute it!

Solutions:

1)To obtain the vis viva equation we proceed as follows:

(a) subtract $C = \frac{1}{2}V_P^2 - \mu/a(1 - e)$ from: $\frac{1}{2}V^2 - \mu/r = C$

E.g.

$\frac{1}{2}V^2 - \mu/r - \frac{1}{2}V_P^2 - \mu/a(1 - e)$

$= \frac{1}{2}V^2 - \frac{1}{2}V_P^2 - \mu/r - \mu/a(1 - e)$

Then, substitute in: $(V_P)^2 = \mu/a\ [(1 + e)/\ (1 - e)]$ to eliminate V_P, viz.

$\frac{1}{2}V^2 - \frac{1}{2}\ [\mu/a\ [(1 + e)/\ (1 - e)]] - \mu/r - \mu/a(1 - e)$

With appropriate algebraic simplification, yields:

115

$V^2 = \mu (2/r - 1/a)$

2) a) The key here is to recognize "two birds" can be killed with one stone, that is, obtaining the velocities for this section while also obtaining h for part b. This in turn depends on using specific algebraic properties to express h in terms of μ, a and e. In so doing we get:

$h = \pm [\mu a(1 - e^2)]^{1/2}$

where $\mu = 1.33 \times 10^{20}$ Nm²/kg

(Note: for μ, we already know G and m1 = 1.99×10^{30} kg (sun's mass) and m2 = 6.4×10^{24} kg, Earth's mass)

Also: a = 1.496×10^{11} m

Then h = 4.46×10^{15} N-m/kg = 4.46×10^{15} J/kg

The velocity *at perihelion* is then:

$V_P = h/ a(1- e) =$

4.46×10^{15} J/kg / [1.496×10^{11} m(1 - 0.016)]

$V_P = 3.03 \times 10^4$ m/s

and the velocity at aphelion:

$V_A = h/ a(1 + e) =$

4.46×10^{15} J/kg / [1.496×10^{11} m(1 + 0.016)]

$V_A = 2.93 \times 10^4$ m/s

b) We need the energy constants C(A) and C(P) at each of these points, and h, the 'specific relative angular momentum'

We already obtained h, in (a) so need only find the energy constants. We do so for each of the points, perihelion and aphelion.

Then:

$C(P) = \frac{1}{2} V_P^2 - \mu/[a(1-e)]$

$C(P) = \frac{1}{2}\{3.03 \times 10^4 \text{ m/s}\}^2 - (1.33 \times 10^{20} \text{ Nm}^2/\text{kg}) / [1.496 \times 10^{11} \text{ m}(1 - 0.016)]$

$C(P) = -4.45 \times 10^8 \text{ m}^2/\text{s}^2$

$C(A) = \frac{1}{2} V_A^2 - \mu/[a(1+e)]$

$C(A) = \frac{1}{2}\{2.93 \times 10^4 \text{ m/s}\}^2 -$

$(1.33 \times 10^{20} \text{ Nm}^2/\text{kg}) / [1.496 \times 10^{11} \text{ m}(1 + 0.016)]$

$C(A) = -4.45 \times 10^8 \text{ m}^2/\text{s}^2$

And not surprisingly, we see they are the same (energy) constants.

c) To confirm the results obtained in (a):

Vis viva states:

$V^2 = \mu (2/r - 1/a)$ or $V = [\mu (2/r - 1/a)]^{1/2}$

If it is to confirm the results in (a) then it should give *the same velocities* when: r1 (perihelion radius vector)

$= 0.98329$ AU $= (0.98329)(1.496 \times 10^{11} m)$

Or: $r1 = 1.47 \times 10^{11}$ m

And, r2(aphelion radius vector) =

1.01671 AU $= (1.01671) (1.496 \times 10^{11} m)$

Or: $r2 = 1.52 \times 10^{11}$ m

Then, call V1 the velocity at r1 (e.g. perihelion):

$V1 = [\mu (2/r1 - 1/a)]^{1/2}$

$V1 = [(1.33 \times 10^{20} Nm^2/kg)[2/1.47 \times 10^{11} m - 1/1.496 \times 10^{11} m]^{1/2}$

$V1 = 3.03 \times 10^4$ m/s which is the same as V_P obtained in (a). Similarly:

$V2 = [\mu (2/r2 - 1/a)]^{1/2}$

$V2 = [(1.33 \times 10^{20} Nm^2/kg)[2/1.52 \times 10^{11} m - 1/1.496 \times 10^{11} m]^{1/2}$

$V2 = 2.93 \times 10^4$ m/s or the same as V_A obtained in (a). The vis viva equation therefore confirms the earlier results.

3) To improve the value for Jupiter's mass (from what the first problem yields) using: $T = 2\pi (a^3/\mu)^{1/2}$ (Recall $\mu = G(m_1 + m_2)$) and Jupiter's mass and the Sun's have already been given along with G, in the general information)

We have:

$m_1 = 1.99 \times 10^{30}$ kg

But the key in this problem is to use $T = 2\pi (a^3/\mu)^{1/2}$ to obtain μ, and thence m_2 for Jupiter as the sole unknown. This means first arranging the equation to make $T = 2\pi (a^3/\mu)^{1/2}$ the subject, viz.

$T^2 = 4\pi^2 (a^3/\mu)$ (squaring both sides)

$T^2 / 4\pi^2 = (a^3/\mu)$

or:

$\mu T^2 = 4\pi^2 a^3$ (cross multiplying)

Then:

$\mu = G(m_1 + m_2) = 4\pi^2 a^3 / T^2$

Now, $a = 5.2$ AU $= 7.77 \times 10^{11}$ m

And the period of Jupiter (from previous blogs) is $T = 11.86$ yrs. $= 3.74 \times 10^8$ s

Then:

$4\pi^2 a^3 / T^2 = (4\pi^2)[7.77 \times 10^{11} \text{ m}]^3 / [3.74 \times 10^8 \text{s}]^2$

So:

$G(m_1 + m_2) = 1.32 \times 10^{20}$ Nm²/kg

whence (since we note $m_1 \gg m_2$):

$(m_1 - m_2) = 1.32 \times 10^{20}$ Nm²/kg/ G

or:

$(1.99 \times 10^{30}$ kg $-m_2) =$

1.328×10^{20} Nm²/kg/ $(6.7 \times 10^{-11}$ Nm²/kg²)

$(1.99 \times 10^{30}$ kg $-m_2) \approx 1.981 \times 10^{30}$ kg

$m_2 = 1.99 \times 10^{30}$ kg $- 1.981 \times 10^{30}$ kg $= 0.009 \times 10^{30}$ kg $= 9 \times 10^{27}$ kg

(compared with the value of 1.89×10^{27} kg in tables.)

4) The approach to solution is analogous to that we used in solving simpler astronomy problems in terms of using a pair of ratios to apply to Kepler's 3rd law, viz.

$(P_1/ P_2)^2 = k(a_1/ a_2)^3$

In this case we use the key equation at the end of Part (6) which relates the period T, to $\mu = G(m_1 + m_2)$.

$T = 2\pi (a^3/\mu)^{1/2}$

Now, in this problem we are dealing with Jupiter (planet) and its 5th satellite, with all the respective parameters given. Then we will have to pair two different ratios in a, T, i.e. of the satellite (moving around Jupiter) to Jupiter (around the Sun). We simply apply the same type of approach as shown with P1, P2 to a1, a2, thus:

For the planet and Sun:

$T = 2\pi (a^3/\mu)^{1/2}$

For the planet and satellite (Jupiter and its 5th satellite):

$T' = 2\pi [(a')^3/\mu']^{1/2}$

Then, dividing the bottom form by the top:

$T'/T = \{[(a')^3/ a^3] [\mu/\mu']\}^{1/2}$

Or:

$\mu/\mu' = (T'/T)^2 (a/a')^3$

where $\mu = (M + m)$ and $\mu' = (m + m')$

Given G (Newtonian gravitational constant) cancels out, M= Sun's mass, m = Jupiter's mass, and m' = satellite's mass.

This is simplified by reducing a, T to AU and yrs, so that: a = 5.203 AU, a' = 0.001207 AU, T = 11.86 yrs. and T' = (0. 4982/365.25) yr = 0.00136 yr. Then:

$(M + m)/(m + m') =$

$(0.00136/11.86)^2 (5.203/0.001207)^3$

Now the left side can be posed (by appropriate algebraic manipulation):

$(M + m)/(m + m') = (M/m) [(1 + m/M)/(1 + m'/m)]$

and in the limit of $m/M >>> 1$ we can write:

$M/m = (0.00136/11.86)^2 (5.203/0.001207)^3$
$= 1.059 \times 10^3$

or: $M = (1.059 \times 10^3) m$

So, the mass of the Sun is (1.059×10^3 Jupiter's) or inverting:

$m/M = 1/(1.059 \times 10^3) = 9.438 \times 10^{-4}$

The ratio of Jupiter's mass to that of the Sun.

We can check this accuracy by noting the mass of the Sun is given as 1.99×10^{30} kg, so the mass of Jupiter would be:

$m = (9.438 \times 10^{-4}) (1.99 \times 10^{30}$ kg$) = 1.878 \times 10^{27}$ kg

which compares to the (tabulated) mass of 1.89×10^{27} kg (from sample problem (1).) We thereby obtain a much higher accuracy, likely from using a ratio of *two compounded masses* and the refinement provided by Newton's version of Kepler's 3rd law.

5) We know communications satellites in circular equatorial orbits about Earth always remains above a point of fixed longitude. Clearly, the same exact approach obtains here as we saw for (4), the only difference is we're now considering the Earth-artificial satellite as a separate system in relation to the Earth and Sun - but their properties (a, T) will be similarly paired from Newton's version of the 3rd law of Kepler. Thus again:

$\mu/\mu' = (T'/T)^2 (a/a')^3$

where $\mu = (M + m)$ and $' = (m + m')$

Given G (Newtonian gravitational constant) cancels out, M= Sun's mass, m = Earth's mass, and m' = artificial satellite's mass.

The satellite period T' = 23h 56m = 86 160s = (86160)/31 557 600 = 0.0027 yr.

and a' = $(4.18 \times 10^7$ m$)/ (1.495 \times 10^{11}$ m/AU$)$ = 2.79×10^{-4} AU

Then:

$(M + m)/(m + m') = (0.0027/1)^2 (1/2.79 \times 10^{-4})^3$

From (4) we found:

$(M + m)/(m + m') = (M/m) [(1 + m/M)/(1 + m'/m)]$

and in the limit of m/M > > > 1 we can write:

M/m = 3.4×10^5

The ratio of the mass of the Sun to the mass of the Earth.

Then the mass of the Earth will be:

m = (1.99 x 10^{30} kg)/ (3.4 x 10^5) = 5.83 x 10^{24} kg

(Compared to the tabulated value of 5.97 x 10^{24} kg)

6) In this case, we relate the paired orbital parameters (a, T) of Jupiter to Earth, then make the appropriate mass change to solve for the new period. In this case, it helps to begin with the mass sums first.

We have:

μ/μ' = (T'/T)^2 (a/a')^3

where μ = (M + m) and μ' = (M + m'), the μ applied for Jupiter and Sun, and the μ' for Earth (m') and Sun (M). But we are demanding the new mass condition (Jupiter's mass = Earth's) or m = m', so:

μ/μ' = (M + m')/ (M + m') = 1

Therefore:

1 = (T'/T)^2 (a/a')^3

Let T' = 1 yr (Earth), a' = 1 AU (Earth), and we must then solve for the T, for Jupiter given its mass change to be equivalent to Earth's. We retain its semi-major axis, a = 5.203 AU. Then:

$1 = (1/T)^2 (a/1)^3$

Or:

$T^2 = (a)^3$ so $T = [(a)^3]^{1/2} = [5.203^3]^{1/2} = 11.868$ yrs.

This compares to Jupiter's 11.856 yr. tabulated value to 5 significant figures. In other words, there is only a minor error in its period, of about 0.1%.

7) Is *it really* possible from this information to obtain the semi-major axis for this comet? If yes, then proceed to compute it!

Solution:

(a) The ratio is straightforward and the relation was given earlier:

$(V_P/V_A) = (1 + e)/(1 - e)$

since we know e = 0.9673)

$(V_P/V_A) = (1 + 0.9673)/(1 - 0.9673) = 60.162$

(b) This requires at least the semi-major axis be known, and no, there isn't enough to compute it from the information. But most people (at least amateur astronomers for whom these intermediate level problems are no big deal) *know its period is 86 years.* Then, using Kepler's third law:

$a = [(86 \text{ yr})^2]^{1/3} = 19.48$ AU

This is then enough information to give V_P and V_A at

least in terms of h (the specific relative angular momentum'). We need the actual mass of Halley's comet to obtain h itself.

So we have:

$V_P = h/ a(1 - e) = h /[19.48 (1 - 0.9673)]$

$V_P = h/0.636 = 1.57\, h$

And:

$V_A = h/ a (1 + e)$

$V_A = h/ [19.48 (1 + 0.9673)] = h/38.3 = 0.026\, h$

XI. Binary Star Orbits

We now come to the binary stars which, because they comprise two stars in a dynamically associated system, can be analyzed using the same principles of Kepler's 3rd law used for the planets. Binaries are important in their own right, because they are exactly the systems needed for us to find and corroborate the masses of stars, and hence also test basic theories at the foundation of astrophysics, such as stellar evolution.

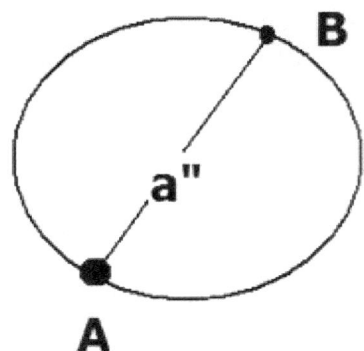

Binary star system with masses m(A) and m(B) with an angular separation a" seconds of arc.

The diagram above shows the simplest type of system, with binary components A (of mass m(A)) and B (of mass m(B)). This visual binary system, it's important to note, indicates the apparent relative orbit of the 2 stars - since ordinarily the plane of the real orbit will not lie in the plane of the sky (that is, perpendicular to the line of sight as portrayed). Usually, what we do is observe the motions of the fainter member about the brighter over a period of time sufficient to determine the orbit period, and then obtain the apparent relative orbit.

Geometrically, it is straightforward to show that an ellipse in one plane when projected onto another plane (say oblique to it) will yield another ellipse but of different eccentricity, $e = c/a$. Most importantly, the foci of the original ellipse do not project onto the foci of the projected one. This means the primary (brightest star) though it is located at one focus of the true relative orbit, is not at the focus of the apparent relative orbit.

But it is this circumstance which makes it possible to determine the inclination of the true orbit to the plane of the sky. Basically, the problem reduces to finding the angle at which the true relative orbit must be projected in order to account for the amount of displacement of the primary from the focus of the apparent relative orbit.

If the semi-major axis of the true relative orbit (e.g. the one it would have if displayed face-on) has an angular distance of a" (seconds of arc) and if the system is at a distance d parsecs, then the semi-major axis in astronomical units is:

$a = (a" \times d)$

Then the sum of the masses of the two stars is given by Kepler's law:

$m(A) + m(B) = (d\, a")^3 / P^2$

where P is their period.

Of course, obtaining the total mass is only the first step. One then wishes to obtain the individual masses for each star. This is done by analyzing the motions of each member with respect to the center of mass of the system which ordinarily will be apart from either member. For example, if m(A) = m(B) then the center of mass (cm) will be exactly midway between them.

Example Problem:

Consider the visual binary system, Sirius A and B. The semi-major axis of the true relative orbit is 7½" and the distance from the Sun to Sirius is 2.67 pc. If the period of the Sirius binary system is 50 years and the component B is found to be twice as far from the center of mass as component A, then **find the total mass of the Sirius system and the masses of each component**.

Solution:

We first obtain the mass total:

m(A) + m(B) = (d a")^3/ P^2 = (2.67 pc x 7.5")/(50 yr)^2 = 3.2 solar masses

So:

m(A) + m(B) = 3.2 M$_s$

But in terms of the center of mass:

A O------------x cm-------------------------o B

where: xB = 2 (xA)

Then: $x_B/x_A = 2$, and: $m(A)/m(B) = x_B/x_A = 2/1$

so: $m(B) = \frac{1}{2} m(A)$

(since the more massive star is always closer to the center of mass)

Thus, $m(A) \approx 2.13\ M_s$, and $m(B) \approx 1.06\ M_s$

Spectroscopic binaries are also of much interest and derive their name because spectroscopic analysis is needed to obtain the radial velocity curve for the system and hence the relative orbital velocity, V, for the pair. The distance around the relative orbit, its circumference, is just the relative orbital velocity V (deduced from the radial velocity profile) multiplied by the period. Then the distance between the stars a, is just:

$a = (V \times P)/ 2\pi$

If, for instance, the relative velocity is a lower limit, then the separation we obtain is a lower limit for the system. If this is then applied to Kepler's 3rd law one can obtain a lower limit to the sum of the masses of the components:

$m_1 + m_2 = a^3/P^2$

Example Problem:

A spectroscopic binary system is found to have a relative velocity of 100 km/sec and a period of 17.5 days. Obtain a lower limit for the separation of the components, a, and thence a lower limit to the sum of the masses. If the spectroscopic analysis shows component (1) is 3 times the mass of component (2), find the lower limits on the masses of the components.

Solution:

First convert 100 km/sec to AU/yr.

Over one year: $t = 3.156 \times 10^7$ s

total distance covered:

d = v x t = 100 km/s (3.156 x 10 ^{7} s) = 3.156 x 10^9 km

But 1 AU = 1.495 x 10^8 km

Then, the AU in this total distance:

d/AU = (3.156 x 10^9 km)/ (1.495 x 10^8 km) = 21.1 AU

The period in yrs. for 17.5 days:

P = 17.5/ (365.25) = 0.048 yr.

Then:

a = (21.1 AU x 0.048 yr/AU)/ 2π = 0.161 AU

The lower limit to the masses is therefore:

m1 + m2 = $(0.161)^3$/ $(0.048)^2$ = 1.8 solar masses

since star m1 has 3x the mass of star m2, then:
$m_1 = 3m_2$

And: $m_2 + 3m_2 = 1.8$ **or** $4m_2 = 1.8$

so: $m_2 = 1.8/4 = 0.45$ solar masses, and $m_1 = 3(0.45) = 1.35$ solar masses

Other Problems:

(1) Find, approximately, the periods of revolution of the following binary star systems in which each star has the same mass as the Sun, and in which the semi-major axis of the relative orbits has the value:

(a) 1 AU

(b) 6 AU

(c) 100 AU

(2) For each of the systems in (1), at what distance would the two stars appear to have an angular separation of 1"?

(3) The true relative orbit of Epsilon Ursae Majoris has a semi-major axis of 2½" and the parallax of the system is 0."127. If its period is 60 years, find the sum of the components in solar mass units.

(4) A hypothetical spectroscopic-eclipsing binary system is observed and its period is 3 years. The maximum radial velocities with respect to the center of mass of the system are:

Star A: $4\pi/3$ AU/yr and Star B: $2\pi/3$ AU/yr

(a) Find the ratio of the masses of the components.

(b) Find the mass of each star in solar units.

(Assume the eclipses are central)

Solutions:

   (1) For each system we have: $m_1 = m_2$ so total mass = $2M_s$ (solar masses)

Then: $m_1 + m_2 = (a)^3 / P^2$ and

$P = [(a^3)/(2)]^{1/2}$

(a) IF a = 1 AU

$P = [1/2]^{1/2} = 0.707$ yr.

(b) IF a = 6 AU

$P = [(6)^3 / 2]^{1/2} = 10.3$ yr.

(c) IF a = 100 AU

$P = [(100)^3/2]^{1/2} = 707.1$ yrs.

(2) We need to determine the distances for the two stars in each system of (1) to appear to have an angular separation of 1".

We require:

$m_1 + m_2 = (d\ a")^3 / P^2$

and need to find d for different a" = 1"

Simplifying:

(a) P = 0.707 yr.

$(d\ a") = [(m_1 + m_2)\ P^2]^{1/3}$

Then: $d = [(2)(0.707)^2]^{1/3} = 1$ pc

(b) P = 10.39 yr.

Then: $d = [(2)(10.3)^2]^{1/3} = 6$ pc

c) P = 707.1 yr

Then: $d = [(2)(707.1)^2]^{1/3} = 100$ pc

(3) We note the *true relative orbit* of Epsilon Ursae Majoris has a semi-major axis of 2½" and the parallax of the system is 0."127. The period of 60 years then allows us to find the sum of the components in solar mass units.

First, find the distance d from the parallax method:

d = 1/p" = 1/ 0."127 = 7.87 pc

Apply Kepler's 3rd law for binaries separated by a".

$m(A) + m(B) = (d\ a")^3 / P^2$

where a" = 2½" and d = 7.87 pc with P = 60 yrs.

Then:

m(A) + m(B) = ((7.87) (2½")]³/ 60² = 2.1 solar masses

4) The graphs of the radial velocity curves for the stars, matching the physical situation of each, is given in the diagram below. The important thing is to have the maxima and minima in the correct directions at the key points in their respective orbits.

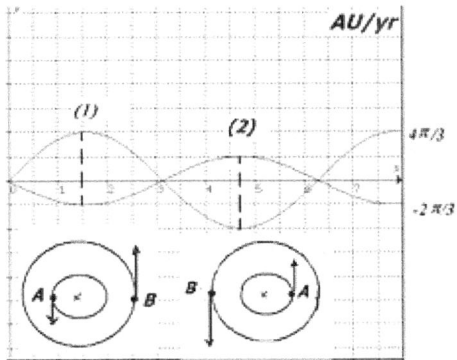

The radial velocity curves and depictions of key situation for problem #4:
At position (1), star A has its max. velocity component toward the Sun, and star A its max. velocity away. The conditions are reversed at position (2). Since the stars are moving in opposite directions relative to c.m. the max. radial velocity = 2π AU/yr

Positions (1) and (2) in the graphs allow us to obtain the maximum relative velocity for the system, for which:

V = (4π/3 + 2π/3)AU/yr = 6π/3 AU/yr = **2π AU/yr**

P = 3 yrs.

(a) Find the ratio of the masses of the components.

The ratio of the masses will be in the ratio of the radial velocities or:

$4\pi/3 : 2\pi/3 = 4\pi/2\pi = 2:1$

(b) To get the mass of each star in solar units. (Assume the eclipses are central)

Recall that to get distance:

$a = (V \times P)/2\pi = (2\pi \text{ AU/yr} \times 3 \text{ yr})/2\pi = 3 \text{ AU}$

and: $m_1 + m_2 = (3\text{AU})^3/(3 \text{ yr})^2 = 3$ solar masses

The masses are *inversely proportional to the radial velocities* in their orbits. Since:

Star A has $v_A = 4\pi/3$ AU/yr $= 2 v_B$ where $v_B = 2\pi/3$ AU/yr

then: $m(A)/m(B) = v_B/v_A = 1/2$

or $m(A) = m(B)/2$ or $m(B) = 2m(A)$

But: $m(A) + m(B) = 3$ so: $m(A) + 2m(A) = 3$

$3 m(A) = 3$

and $m(A) = 1$ solar mass, $m(B) = 2m(A)$ or, 2 solar masses

XII. Introducing Spherical Astronomy

Practical astronomy, including astrometry, is a vast sub-discipline of astronomy and it is also among the richest, most rewarding fields. As the name implies, it entails learning about the mechanics of the sky: how to measure angles and reference coordinates, then how to use these to find astronomical objects in terms of their positions, including altitude for the observer, as well as azimuth.

But before one can do all those things, one has to become familiar with the sky coordinate system and geometry, ultimately working with the basic relations for spherical trigonometry. This is merely an extension of plane trig, but to the sort of angles (many > 90 degrees) one finds in spherical or astronomical applications.

Fig. 1

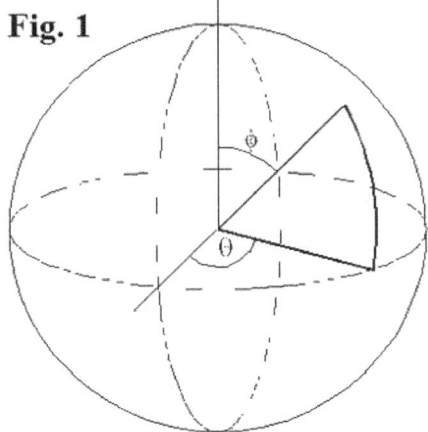

SPHERICAL GEOMETRY

A simple illustration of a spherical geometry is shown in Fig. 1. In the diagram, the angle Θ denotes the longitude measured from some defined meridian on the sphere, while the angle φ denotes a zenith distance, or the measured angle from an object to the zenith.

Fig. 2

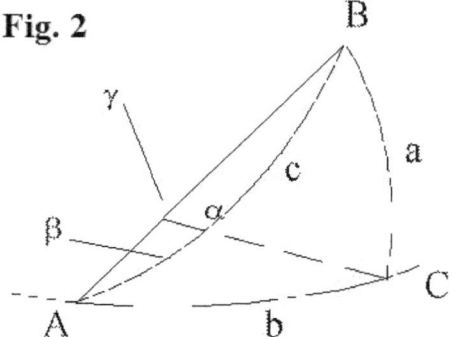

SPHERICAL RIGHT TRIANGLE

Fig. 2 shows a spherical right triangle from which a host of different angle relationships can be obtained, which can then be used to find astronomical measurements, etc.

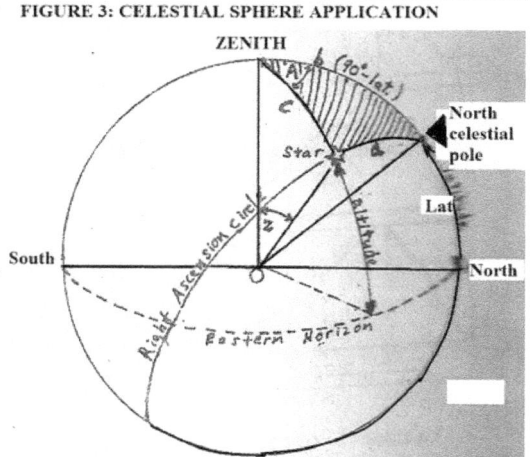

FIGURE 3: CELESTIAL SPHERE APPLICATION

Fig.3 shows an actual example of a celestial sphere, such as used in many practical astronomy applications, and some of the key angles with reference to a particular object (star) referenced within a given coordinate system. In some applications, the coordinate system may not need to be changed, but in others it must - for example, when going from the coordinate system applied to sky objects (Right Ascension, Declination) to the observer's own coordinates (altitude, azimuth). In this way, coordinate transformations will also enter and we'll get to those in time.

For now, let's consider a simple angle relation in Fig. 1, to find the altitude, a. Then if we have the basic geometrical relationship: a + φ = 90 degrees, clearly then a = (90 - φ).

Let's now examine Fig. 2 and see what spherical trig relationships we can infer.

Two of the key ones embody the law of sines and law of cosines for spherical triangles, which are the analogs of the law of sines and cosines in plane trig.

We have for ***the law of sines***:

sin A/ sin a = sin B/ sin b = sin C/ sin c

where A, B, C denote ANGLES and a,b,c denote measured arcs. (Note: we could also have written these by flipping the numerators and denominators).

We have for the law of cosines:

cos a = cos b cos c + sin b sin c cos A

Where a, b, c have the same meanings, and of course, we could write the same relationship out for any included angle.

Now, we use Fig. 3, for a celestial sphere application, in which we use the spherical trig relations to obtain an astronomical measurement.

Using the angles shown in Fig. 3 each of the angles for the law of cosines (given above) can be found. They are as follows:

cos a = cos (90° - δ)

where δ = declination

cos b = cos (90° - Lat)

where 'Lat' denotes the latitude. (Recall from Fig. 1 if φ is polar distance (which can also be zenith distance) then φ = (90 - Lat))

cos c = cos z

where z here is the zenith distance.

sin b = sin (90 deg - Lat)

sin c = sin z

and finally,

cos A = cos A

Where A is the *azimuth*.

Example Problem:

Let's say we want to find the declination of the star if the observer's latitude is 45° N, the azimuth of the star is measured to be 60°, and its zenith distance z = 30°. Then one would solve for cos a:

$\cos a = \cos(90° - \delta) =$

$\cos(90° - \text{Lat}) \cos z + \sin(90° - \text{Lat}) \sin z \cos(A)$

$\cos(90° - \delta) = \cos(90° - 45°) \cos 30°$

$+ \sin(90° - 45°) \sin 30° \cos 60°$

And:

$\cos(90° - \delta) = \cos(45°) \cos 30°$

$+ \sin(45°) \sin 30° \cos 60°$

We know, or can use tables or calculator to find:

$\cos 45° = \sqrt{2}/2$

$\cos 30° = \sqrt{3}/2$

$\sin 45° = \sqrt{2}/2$

$\sin 30° = 1/2$

$\cos 60° = 1/2$

Then:

cos (90 - δ)= {(√2/ 2)(√3/ 2)} + {√2/ 2} √ (½) }

cos (90 - δ)= √6/ 4 + √2/ 8 = {2√6 + √2}/ 8

cos (90 - δ) = 0.789

arc cos (90 - δ)= 37.°9

Then:

δ = 90 ° - 37.° 9 = 52.° 1

Or, in more technical terms:

δ (star) = + 52.1 degrees

Problem:

The nearest planet to the Sun, Mercury, is due to make its brightest appearance on April 1. It will have a Right Ascension of 1 h 24 m and a declination of 12 deg 27 '.

If an observer (located at latitude 40 deg north) wishes to find it, where would he look at about 6.45 p.m.? Give the azimuth, A and the altitude a.

Solution:

First, one needs to obtain the sidereal time. From a table of sidereal times this is ~ 7h 25 m. (For an observer at 40 N).

Next, we change the Right Ascension to hour angle

using: h = S.T. - RA

So: h = 7h 25m - 1h 24m = 6h 01m

This is then converted into degrees, using the fact that there are 15 degrees/ hr.

So: 6h 01 m ≈ 90 degrees   (e.g. 6h x 15 deg/ h)

We then find the zenith distance, z, using an astronomical triangle using sides with hour angle, declination and latitude. This implies the spherical version of the law of cosines, and by analogy with the previous solution we gave.

we have:

cos z = sin (decl.) sin (Lat) + cos (decl.) cos h cos (Lat)

We note the following respective values:

sin (δ) = sin 12°.45 = 0.2155

sin (Lat) = sin 40 = 0.6427

cos (δ) = cos 12°.45 = 0.9764

cos h = cos 90° = 0

so, effectively:

cos z = sin (δ) sin (Lat) = (0.2155)(0.6427) = 0.1385

Thence:

$z = \arccos(0.1385) = 82°$

Then, a (altitude) $= 90° - z = 90° - 82° = 8°$

Which shows that we need to look about 8° above the horizon.

But in *which direction*? This requires the azimuth:

The appropriate astronomic triangle yields the following equation for A, azimuth:

$\tan A = -\cos(\delta) \sin h / [\sin(\delta)\cos(\text{Lat})$

$- \cos(\delta) \cos h \sin(\text{Lat})]$

The additional values we need to what was given above are:

$\sin h = \sin 90° = 1$

$\cos(\text{Lat}) = \cos 40° = 0.7660$

Then the computation is displayed as follows:

$\tan A = -(0.9764) / \{(0.2155)(0.7660)\}$

since the 2nd term in the denominator drops out, as $\cos h = \cos 90° = 0$

then:

$\tan A = -(0.9764) / (0.1650) = -5.914$

Then:

A = arc tan (-5.914) = -80°.4

Since the value is negative, we take:

A = 360° - 80°.4 = 279°.5

which pins it just about at the azimuth shown in the diagram below:

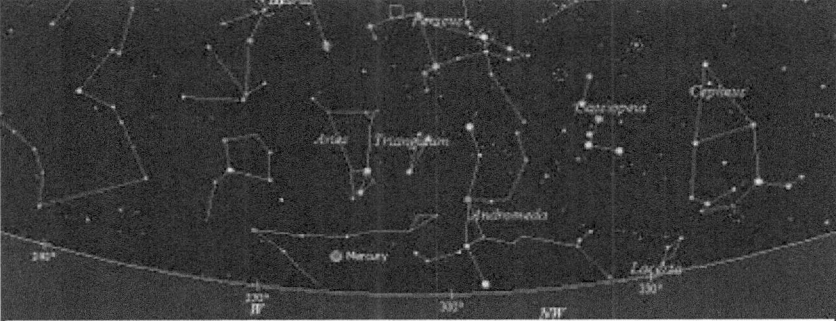

FIG. 1: The visible sky on April 1, 2011, with view toward Western horizon and Mercury's position just over 280 deg azim.

A similar problem:

You are located in San Francisco, and the sidereal time at 11.45 p.m. local time for your location on this date, is 14 h 15 m, approximately. Saturn is visible and is at 12h 47 m Right Ascension, and at a Declination of (- 2° 10').

If your latitude is 37° 46 ½' north, find where Saturn will be visible in terms of its altitude and azimuth.

Solution:

The sidereal time (S.T.) is given as: 14 h 15 m

Next, we change the Right Ascension (12h 47m) to hour angle using:

h = S.T. - RA

So: h = 14 h 15 m − 12h 47 m = 1 h 28 m,

This is then converted into degrees, using the fact that there are 15 degrees/ hr.

So: 1h 28m ≈ 22°.5 (e.g. 3h/2 x 15°/ h = 22°.5)

We then find the zenith distance, z, using an astronomical triangle using sides with hour angle, declination and latitude. This implies the spherical version of the law of cosines, and by analogy with the previous solution we have:

cos z = sin (δ) sin (Lat) + cos (δ) cos h cos (Lat)

We note the following respective values:

sin (δ) = sin (-2°.16) = -0.0376

sin (Lat) = sin (37° 46 ½') = sin 37°.77 = 0.6124

sin h = sin 22°.5 = 0.3826

cos (δ) = cos (-2°.16) = 0.9992

cos h = cos (22°.5) = 0.9238

cos(Lat) = cos 37°.77 = 0.7904

so, effectively:

cos z = sin (-2°.16) sin (37°.77)
+ cos (-2°.16) (cos 22°.5) (cos 37°.77)

cos z = (-0.0376)(0.6124) +(0.9992)(0.9238)(0.7904)

cos z = -0.023 + 0. 7295 = 0.7065

z = arc cos (0.7065) = 45.° 0

Then,

a (altitude) = 90° - z = 90° - 45° = 45°

Which shows that we need to look about halfway up above the horizon, toward the zenith.

But in which direction?

This requires the azimuth, A:

The appropriate astronomic triangle yields the following equation for A, azimuth:

tan A = -cos(δ) sin h / [sin(δ)cos(Lat) –

cos(δ) cos h sin (Lat)]

Take the numerator first and compute it:

$-\cos(-2.16)(\sin 22.5) = -(0.9992)(0.3826) = -0.3822$
Then the denominator:

$\sin(-2°.16)\cos(37°.77) -$

$\cos(-2°.16)\cos 22°.5 \sin(37°.77)$

$= (-0.0376)(0.7904)) - (0.9992)(0.9238)(.6124)$

$= -0.0297 - 0.5652 = -0.5949$

so that:

$\tan(A) = (-0.3822)/(-0.5949)$

$\tan(A) = 0.6424$

$A = \arctan(0.6424) = 32°.7$

This is positive, so this angle is added to the azimuth for the south point of the horizon (180°) to get:

A (Saturn) = 180° + 32.°7 = 212.°7 , or very near to the actual value (see star map for time and date) of 212.° 6 (or *212 degrees 36 mins.*, i.e. 32 degrees and 36 mins. **west of the south point of the horizon**)

The diagrammatic solution is shown below, based on inputting the computed values to the *Cybersky* astronomy program:

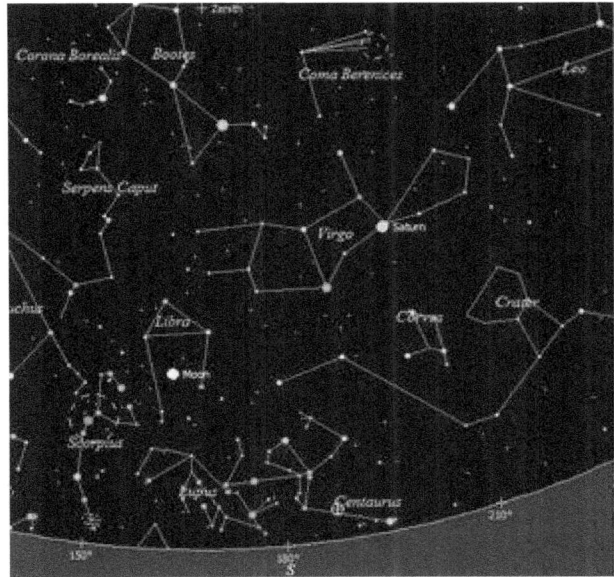

The celestial map for the date in question at San Francisco - the actual altitude of Saturn is a = 45 deg 06 ' and the azimuth A = 212 deg 36'= 212.6 deg or very near to what we obtained via astronomical triangle.

Of course, one can note from these horizon coordinates, using separate computations, that the date in question is May 16, 2011.

We will now, for completeness, consider a very powerful and efficient method to solve coordinate transformation problems which makes use of matrices.

2. Matrix Methods

In the previous section we examined how to obtain horizontal coordinates for celestial objects by using the so-called "astronomical triangle". (See diagram below):

CELESTIAL SPHERE for Matrix Applications

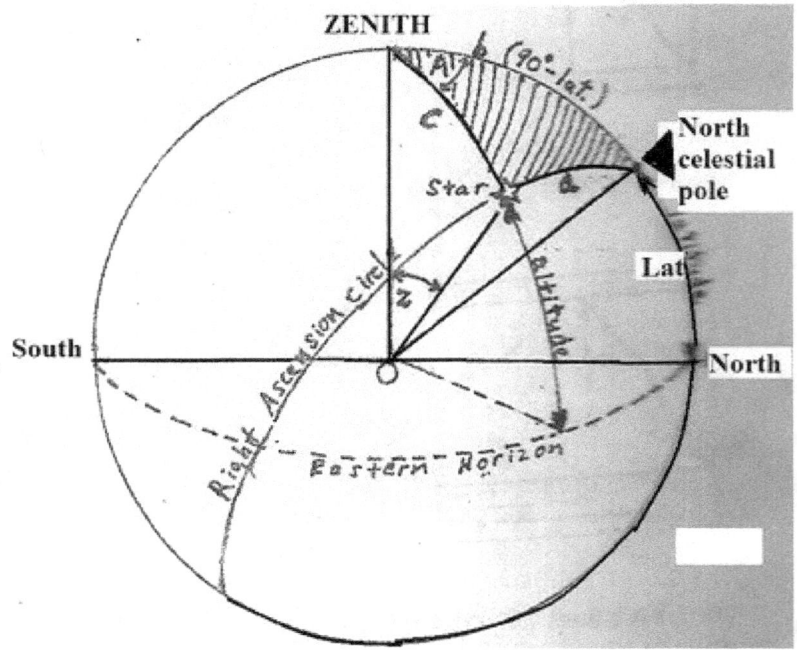

    There is, however, an alternative method which makes use of matrices - mainly of the 3 x 3 variety. In this blog we explore this other method, and show how it can be used.

    The basic principle involves relating **the Cartesian coordinates (rectilinear)** of a point on the celestial sphere (diagram) **to the curvilinear coordinates** measured in the primary and secondary reference planes. One has then, for example, the left side showing the Cartesian x, y, z coordinates and the right side the curvilinear in u and v:

$$\begin{pmatrix} x \\ y \\ z \end{pmatrix} := \begin{pmatrix} \cos(v)\cos(u) \\ \cos(v)\sin(u) \\ \sin(v) \quad 0 \end{pmatrix}$$

After conversion the curvilinear coordinates may be calculated according to:

u = arctan (y/x) and v = arcsin (z)

Consider conventional orthogonal matrices of 3 x 3 dimensions, given as functions: R1(Θ), R2(Θ) and R3(Θ), to rotate the general system by the angle Θ about axes x, y and z, respectively. Thus we obtain:

$$R_{1\theta} := \begin{pmatrix} 1 & 0 & 0 \\ 0 & \cos(\theta) & \sin(\theta) \\ 0 & -\sin(\theta) & \cos(\theta) \end{pmatrix}$$

$$R_{2\theta} := \begin{pmatrix} \cos(\theta) & 0 & -\sin(\theta) \\ 0 & 1 & 0 \\ \sin(\theta) & 0 & \cos(\theta) \end{pmatrix}$$

And finally:

$$R_{3\theta} := \begin{pmatrix} \cos(\theta) & \sin(\theta) & 0 \\ -\sin(\theta) & \cos(\theta) & 0 \\ 0 & 0 & 0 \end{pmatrix}$$

Now, I invite readers to use the example of the Mercury problem to apply the matrix methods.

To arrive at the same values (A, a) we will require:

$R_3(\Theta) = R_3(-180°)$

$R_2(\Theta) = R_2(90° - \text{lat.})$

so that:

(x)
(y)
(z) A,a = **$R_3(-180°)$ $R_2(90 - \text{lat.})$ $(XYZ(h, \delta))$**

where : $(XYZ(h, \delta))$ =

(x)
(y)
(z) h, δ

To get you started, we have:

$$R_3(180) := \begin{pmatrix} \cos(180) & \sin(180) & 0 \\ -\sin(180) & \cos(180) & 0 \\ 0 & 0 & 0 \end{pmatrix}$$

Since we saw:

$$R_{3\theta} := \begin{pmatrix} \cos(\theta) & \sin(\theta) & 0 \\ -\sin(\theta) & \cos(\theta) & 0 \\ 0 & 0 & 0 \end{pmatrix}$$

Therefore: $R_3(-180) =$

$$\begin{pmatrix} -1 & 0 & 0 \\ 0 & -1 & 0 \\ 0 & 0 & 1 \end{pmatrix}$$

Obtaining $R_2(\Theta) = R_2(90° - \text{lat.})$ is just as easy, if one recalls the basic trig identity:

$$\cos(90° - \varphi) = \sin(\varphi)$$

Applying this:

$R_2(90° - \text{lat.}) =$

$$\begin{pmatrix} \sin\text{lat.} & 0 & -\cos\text{lat.} \\ 0 & 1 & 0 \\ \cos\text{lat.} & 0 & \sin\text{lat.} \end{pmatrix}$$

And for which we have:

$\sin(\text{lat.}) = \sin 40° = 0.6427$

$\cos(\text{lat.}) = \cos 40° = 0.7660$

Thence, $R_2(90° - \text{lat.}) =$

$$\begin{pmatrix} 0.6427 & 0 & -0.7660 \\ 0. & 1 & 0 \\ 0.7660 & 0 & 0.6427 \end{pmatrix}$$

Meanwhile:

$$\begin{pmatrix} x \\ y \\ z \end{pmatrix} h, \delta = \begin{pmatrix} \cos \delta & \cos h \\ \cos \delta & \sin h \\ \sin \delta & \end{pmatrix}$$

where:

$$\sin(\delta) = \sin 12°.45 = 0.2155$$

$$\cos(\delta) = \cos 12°.45 = 0.9764$$

$$\cos h = \cos 90° = 0, \text{ and } \sin h = \sin 90° = 1$$

Therefore:

$$\begin{pmatrix} \cos \delta & \cos h \\ \cos \delta & \sin h \\ \sin \delta & \ldots \end{pmatrix} = \begin{pmatrix} 0 \\ 0.9764 \\ 0.2155 \end{pmatrix}$$

So we're now set to perform the matrix multiplication:

$$R_3(-180°) \, R_2(90° - \text{lat.}) \, (XYZ(h, \delta)) = \begin{pmatrix} 0.165 \\ -0.976 \\ 0.139 \end{pmatrix}$$

Thereby the rectilinear coordinates can easily be referenced to curvilinear ones in the horizon system - noting the first and easiest element to find is the altitude, since:

a = arc sin(0.139) = 8° So the altitude is : 8°

Meanwhile, the azimuth A =

arc tan (y/x) = arc tan (-0.976/ 0.165) = -5.91

Therefore: A = arc tan(-5.91) = -80.°4

And, since its' negative, we must subtract from 360 degrees:

A= 360° - 80.°4 = 279.°6

which yields the same result we obtained using the spherical trig triangle!

$$\begin{pmatrix} -1 & 0 & 0 \\ 0 & -1 & 0 \\ 0 & 0 & 1 \end{pmatrix} \begin{pmatrix} 0.6427 & 0 & -0.7660 \\ 0 & 1 & 0 \\ 0.7660 & 0 & 0.6427 \end{pmatrix} \begin{pmatrix} 0 \\ 0.9764 \\ 0.2155 \end{pmatrix} = \begin{pmatrix} 0.165 \\ -0.976 \\ 0.139 \end{pmatrix}$$

The matrix order above conforms to that obtained and shows the final result in the solution column for the matrix: (X,Y,Z)A,a

Other Problems:

1) You are located in Miami, Florida and the sidereal time = 9 h 13 m at 9.30 p.m. local time for this date, approximately. Saturn is visible and is at 13 h 44 m Right Ascension, and at a Declination of (- 7° 56'). If your latitude is 25.°75 north, find Saturn's position in terms of its altitude and azimuth. (*Use both* the astronomical triangle and matrix method to obtain your answer.)

2) Use a *matrix method only* for the same location in problem (1) and for the same sidereal time – but applied to the case of the planet Mars which is also visible at the same local time but at: RA = 10h 28m, and δ = +12° 51'.

3) Use either astronomical triangle or matrix method to determine the altitude and azimuth of both Saturn and Mars, for the same date (*and for Miami's local time*) but for Barbados (Longitude 59° 30', latitude of 13° N.

How would you have worked out the sidereal time and the hour angle, h, for each of the planets?

Use whichever method you didn't use initially, to check the results obtained from the method you did use.

4) Using all or some of the information, data from the preceding problems, estimate the calendar date at which these measurements, observations were made.

Part TWO: Astrophysics

XIII. Stellar Masses and Luminosities

In the last chapter we examined how stellar masses were obtained from binary stars, for example, from their radial velocity curves. In this chapter we go further to show how other stellar properties including luminosity and radius can be obtained, as well as effective (or surface) temperature.

We will start with *The Mass-Luminosity relation*, which enables us to infer the actual brightness of a main sequence star based on its mass, is actually due to data from visual and other binaries. Recall the sample problem we looked at in the previous chapter, viz.

The semi-major axis of the true relative orbit for Sirius A and B is 7½" and the distance from the Sun to Sirius is 2.67 pc. If the period of the Sirius binary system is 50 years and the component B is found to be twice as far from the center of mass as component A, then find the total mass of the Sirius system *and the masses of each component.*

We first obtained the mass total:

m(A) + m(B) = (d a")³/ P² = (2.67 pc x 7.5")/(50 yr)² = 3.2 solar masses

So:

m(A) + m(B) = 3.2 M_s

In terms of the center of mass, we then used the sketch:

A **O**------------x cm-------------------------**o** B

where: xB = 2 (xA)

Then: xB/xA = 2, and:

m(A)/m(B) = xB/xA = 2/1

so: m(B) = ½ m(A)

(since the more massive star is always closer to the center of mass)

Thus, m(A) ~ 2.13 M_s, and m(B) ~ 1.06 M_s

Now the next question is: *Can we obtain the actual brightness of either of these?* The answer is we can, but only for star (A) or Sirius A. This is because the Mass-Luminosity relation *only applies to stars on the Main Sequence,* hence gives the mass in terms of the solar mass (the Sun being used as a 'standard' for the stars on the Main Sequence).

From the Mass-Luminosity relation we have:

L'/L = (M'/M)$^{3.5}$

Or Log (L'/L) = 3.5 Log (M'/M)

where L, M refer to solar values and L', M' to stellar values. In this case, for Sirius A, M' = 2.13 solar masses, so (M'/M) = 2.13, and:

3.5 (Log 2.13) = Log (L'/L) = 3.5 (0.3283) = 1.149

But: antilog (1.149) = 14.09

Or: L' = 14 L approx.

So, Sirius A is about 14 times more luminous than the Sun.

Another example:

The intrinsic brightness (luminosity) of Regulus is greater than the Sun's by a factor 120. Find the approximate mass of Regulus.

Here: L'/L = 120 so Log (120) = 3.5 Log (M'/ M)

and we are seeking to find M' in terms of M.

Log (120) = 2.079 = 3.5 Log (M'/ M)

Or:

0.594 = Log (M'/M)

Taking antilogs of each side:

3.93 = (M'/M) or M' = 3.93 M

Therefore Regulus is approximately 4 times the mass of the Sun.

We already saw the use of simple apparent and absolute magnitudes, but in stellar properties' analysis we need to refine this to deal with "absolute bolometric magnitudes" because the brighter stars (or spectral class O and A mainly) require special "color" corrections usually referred to as "bolometric corrections". This refined system of "bolometric magnitudes" is thereby adjusted so the bolometric corrections are small for stars like the Sun (e.g. G class or later) but large for very hot stars where most of the radiated energy is in the unobservable ultraviolet (UV). The Table below gives bolometric corrections for different temperatures and spectral types.

I. Bolometric Correction and effective Temperature as functions of colour index

B-V	B.C.	$\log T_e$
-0.30	-4.1*	4.65*
-0.25	-3.0*	4.46*
-0.20	-2.3*	4.32*
-0.15	-1.7*	4.21*
-0.10	-1.25	4.13
-0.05	-0.95	4.08
0.00	-0.72	4.03
+0.05	-0.52	3.99
+0.1	-0.38	3.95
+0.2	-0.20	3.91
+0.3	-0.09	3.87
+0.4	-0.02	3.83
+0.5	0.00	3.80
+0.6	-0.03	3.76
+0.7	-0.09	3.73
+0.8	-0.18	3.69
+0.9	-0.29	3.66
+1.0	-0.42	3.63
+1.1	-0.57	3.61
+1.2	-0.74	3.58
+1.3	-0.92	3.56
+1.4	-1.11	3.55
+1.5	-1.3*	3.53*
+1.6	-1.5*	3.52*
+1.7	-1.7*	3.52*
+1.8	-1.9*	3.51

* uncertain

Note the difference (B - V) is the "color index" representing the difference in magnitudes m_B and m_V, e.g. (m_B - m_V) where m_B is the apparent magnitude from a blue filter and m_V is the apparent magnitude from the standard yellow or visual filter - most sensitive to wavelengths near 550 nm.

If the absolute bolometric magnitude (M_{bol}) of a star is known, then its luminosity easily can be found as a function of the Sun's luminosity with the relation:

$$\text{Log}(L'/L) = 0.4(M_{bol} - M_{bol}*)$$

where M_{bol} is for the Sun and $M_{bol}*$ is for the star. Note that any given *absolute visual* magnitude (M_V) can *be changed to an absolute bolometric magnitude* by applying a *bolometric correction* such that:

$$M_{bol}* = M_V* + B.C.$$

Example Problem:

The star Almach (Gamma Andromeda) has $(B - V) = +1.3$ and an apparent visual magnitude $m_V* = 2.16$. What bolometric correction should be applied? Also, find the absolute visual magnitude M_V* and the absolute bolometric magnitude $M_{bol}*$ of the star. How does it compare in luminosity to the Sun? (The distance of Almach is 80 pc.)

Solution:

From the Table provided:

We find $(B - V) = +1.3$ corresponds to $B.C. = -0.92$.

The absolute visual magnitude can be found from the apparent visual magnitude. Thus:

$$(m - M) = (m_V* - M_V*) = 5\log(D) - 5$$

and:

$M_V* = m_V* - 5\log(D) + 5 = 2.16 - 5\log(80) + 5$

$M_V* = 2.16 - 5(1.903) + 5 = -2.36$

The absolute bolometric magnitude is:

$M_{bol}* = M_V* + B.C. = (-2.36) + (-0.92) = -3.28$

The relative luminosity as a function of absolute bolometric magnitude is:

$\log(L'/L) = 0.4(M_{bol} - M_{bol}*) = 0.4(4.63 - (-3.28))$

$\log(L'/L) = 0.4(7.91) = 3.16$

antilog $3.16 = 1445$ so: $L' = 1445\ L$

Other Problems:

(1) The apparent V-band (filter) magnitude of a star is 8.72, and it requires a bolometric correction of -0.48. Find the apparent bolometric magnitude of the star. (Hint: Apparent bolometric magnitudes are obtained in an analogous way to the absolute forms)

(2) A star has a color index (B - V) of +1.0 and its apparent B-band magnitude is 6.4. The corresponding bolometric correction is -0.5. Find the apparent bolometric magnitude of the star.

(3) The star Alhena in the constellation Gemini is at a distance of 30 pc. If it has (B - V) = 0.00, and m_V = +1.93, find the apparent B-band magnitude, m_B.

Also find the absolute visual magnitude and the absolute bolometric magnitude of the star.

Find the luminosity of Alhena in terms of the solar value.

Solutions:

1) $m_V = 8.72$

$m_{bol}{}^* = m_V + B.C. = 8.72 + (-0.48) = 8.24$

Hence, $m_{bol}{}^* = +8.24$

(2) $m_B = 6.4$ and $(B - V) = (m_B - m_V) = 1.0$

Then: $m_V = m_B - 1.0 = 6.4 - 1.0 = 5.4$

$m_{bol}{}^* = m_V + B.C.$

$m_{bol}{}^* = 5.4 + (-0.5) = +4.9$

(3) We have the distance of Alhena of 30 pc. If it has $(B - V) = 0.00$, and $m_V = +1.93$, find the apparent B-band magnitude. Also find the absolute visual magnitude and the absolute bolometric magnitude of the star. Find the luminosity of Alhena in terms of the solar value.

Solution:

$(B - V) = m_B - m_V = 0.00$

Then if $m_V = +1.93$, then $m_B = +1.93$.

The absolute visual magnitude $M_V = m_V - 5 \log D + 5$

$D = 30$ pc, so:

$M_V = 1.93 + 5 \log (30) + 5 = 1.93 - 5(1.477) + 5$

$M_V = 1.93 - 7.39 + 5 = -0.46$

The absolute bolometric magnitude is:

$M_{bol}{}^* = M_V + B.C.$

where B.C. = 0.72 (from the table) so:

$M_{bol}{}^* = -0.46 + (-0.72) = -1.18$

The Luminosity L' is obtained using:

$\log (L'/L) = 0.4 (M_{bol} - M_{bol}{}^*)$

where M_{bol} (Sun's) = 4.63

$\log (L'/L) = 0.4 (4.63 - (-1.18)) = 0.4 (5.81) = 2.32$

Antilog (2.32) = 209 approx.

$L'/L = 209$ and $L' = 209$ L or 209 times the solar luminosity.

XIV. The Stellar Interferometer and Applications

In the last chapter we examined how the simple stellar properties of mass and luminosity (intrinsic brightness) can be obtained. Now, we will extend that to see how a star's radius and effective temperature can be found. Once again, all of these properties have been enabled by work on binary stars - which allows mass and luminosity (for main sequence or stable stars) and then further basic extrapolations from these.

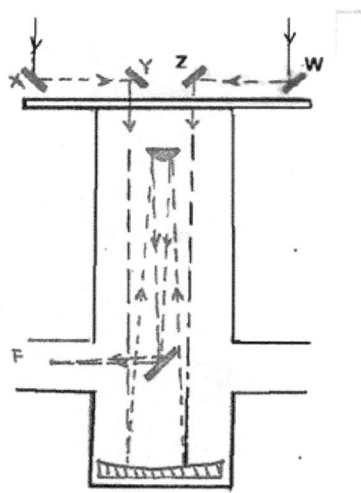

Stellar interferometer: Mirrors x and w are moved inward or out in unison until alternate light and dark fringes are seen in the eyepiece. The angular diam. of the star is then computed when the distance between movable mirrors is known.

Stellar mass and radius essentially come about by relating the star's luminosity (L') to its surface area (A = $4\pi R^2$) and simultaneously to its total energy output (E = $\sigma T'^4_{eff}$), where T'_{eff} is the surface or effective stellar temperature, and σ is the Stefan-Boltzmann constant. When the two are combined, the stellar luminosity can be expressed as:

$L' = 4\pi R'^2 (\sigma T'^4_{eff})$

Or in simple logarithmic form (in terms of the solar values L, T, and R):

$\text{Log}(L'/L) = 4 \text{Log}(T'_{eff}/T_{eff}) + 2 \text{Log}(R'/R)$

Where T_{eff} = 5760 K or the Sun's effective (surface) temperature.

Example Problem:

If the star Sirius A has a mass of approximately twice the Sun's and an effective temperature of 10 000 K, find its radius in terms of the Sun's and also the actual value if the Sun's is R = 6.9 x 10⁵ km.

Solution:

Using the mass luminosity law from before, then if (M'/M) = 2:

$\text{Log}(L'/L) = 3.5 \text{Log}(2) = 3.5 (0.301) = 1.053$

And:

$\text{Log}(L'/L) = 1.053 = 4 \text{Log}(T'_{eff}/T_{eff}) + 2 \text{Log}(R'/R)$

Or, for radius:

$2 \text{Log}(R'/R) = 1.053 - 4 [\text{Log}(T'_{eff} / \text{Log}(T_{eff})]$

Whence:

2 Log (R'/R) = 1.053 - 4 (Log 10000 - Log 5760) = 1.053 - 4(4.00 - 3.76)

2 Log (R'/R) = 1.053 - 0.960 = 0.930

Log (R'/R) = (0.930)/2 = 0.0465

Taking antilogs:

R'/R = 1.11 or R' = 1.11 R

So Sirius A's radius is about 1.1 times the Sun's

Or:

R' = 1.1 (6.9 x 10^5 km) = 7.6 x 10^5 km

It's well to add a cautionary note here that the radius worked out using the preceding method is not the same as that derived from the absolute bolometric magnitude of Sirius.

To show this, note the absolute visual magnitude of Sirius is +1.4 and the bolometric correction corresponding to it is (-0.56). Then the absolute bolometric magnitude is:

M_{bol}* = M_V + B.C. = +1.4 + (-0.56) = +0.84

The luminosity in terms of the Sun's is then:

Log (L'/L) = 0.4 (M_{bol} - M_{bol}*)

Log (L'/L) = 0.4 (4.63 - 0.84) = 1.516

And:

2 Log (R'/R) = 1.516 - 0.960 = 0.556

So:

Log (R'/R) = 0.556/2 = 0.278

Antilog (0.278) = 1.9 (approx.)

Or: R' = 1.9 R

Or, 1.9 times the solar radius.

Which is correct? In fact, both values have uncertainties but the latter value is at least closer to the actual radius, based as it is on more accurate photometric measurements. Besides, the mass-luminosity law (as we saw) is an empirical relationship and also varies over the main sequence (e.g. depending on temperature and luminosity) so we can expect it to differ when used for stars on the upper main sequence, compared to the lower. These refinements, of course, are taken into account in advanced astrophysics- astronomy, but since we are working at an intermediate level, we don't do so here.

Observationally, one method of measuring a star's radius is by using an interferometer such as shown in the sketch. This employs the principle of interference to obtain an angular measurement for the star's apparent diameter (e.g. in arcsec or "). If the distance to the star is known then one can get the radius of the star R* in terms of the Sun's R, by using:

R*/R = (d* a*")/ (d a")

In the foregoing, a" denotes the apparent angular diameter of the sun (1920"), d is the Sun's distance (1 AU = 1.495 x 10^8 km), d* is the star's distance and a*" is the star's apparent angular diameter.

Example Problem:

Find the radius of the star Regulus if its distance is 24 pc and its apparent angular diameter from interferometer measurements is 0."0018.

Solution:

First convert the star's distance to AU to be able to conform with d (= 1 AU)

We know that 200,000 AU = 1 pc, then:

24 pc x (2 x 10^5 AU/pc) = 48 x 10^5 AU

Then:

R*/R = (4.8 x 10^6) x (0."0018/ 1920") = (8640/1920) = 4.5

So: R* = 4.5 R

So Regular has a radius 4½ times that of the Sun.

Other problems:

1) Find the radius of the star Alhena from Problem (1) in the previous problem set. (Use the same data available)

2) Use two different techniques to arrive at the radius of the star Beta Pegasi, which lies at a distance of 50 pc and has: (B - V) = +1.7 and m_v = 2.5. Interferometry obtains the apparent angular diameter as 0."021.

a) What is the apparent discrepancy (as a percentage) in the two values obtained?

b) Account for these discrepancies.

c) Which value, if any, would you be inclined to assign as having greater accuracy?

3) Consider the stars below and their (B- V) and M_v values:

Rigil Kent : (B - V) = 0.71, M_v = 4.2

Spica: (B - V) = -0.23, and M_v = (-3.1)

Fomalhaut: (B - V) = 0.09 and M_v = 1.9

Suhail: (B - V) = 1.7 and M_v = (-4.3)

a) Find the mass, luminosity and radius of each of these stars.

b) Try to estimate the surface temperature of each of these stars. (Hint: you can use the data from the Table of Bolometric corrections in the previous chapter or one of the Tables in the Appendix).

Solutions:

1) To obtain the radius we use:

$2 \log (R^*/R) = \log (L^*/L) - 4 \log (T_{eff}^*/T_{eff})$

Where (from the data in previous problem (1):

$\log (L^*/L) = 0.4 (4.63 - (-1.18)) = 0.4 (5.81) = 2.324$

So:

$2 \log (R^*/R) = 2.324 - 4(4.03 - 3.76) = 1.244$

Or:

$\log (R^*/R) = 1.244/2 = 0.622$ and antilog (0.622) = 4.2

Thus: $(R^*/R) = 4.2$ and $R^* = 4.2$ R or 4.2x the radius of the Sun.

2) Begin by converting 50 parsecs (the distance to Beta Pegasi) to AU:

50 pc = 50 pc x $(2 \times 10^5$ AU/pc$) = 10^7$ AU

We use the relation:

$R^*/R = (d^* a^{*\prime\prime})/ (d\ a^{\prime\prime})$

where: $d^* = 10^7$ AU, $a^{*\prime\prime} = 0.''021$

d = 1 AU, and a" = 1920"

Then:

R*/R = (0."021/ 1920") x (10^7) = 109.3

Thus, by interferometry, R* = 109.3 R

Or, Beta Pegasi is 109.3 times the radius of the Sun.

By photometry:

(B - V) = +1.7 so B.C. = -1.7, while m_v = +2.5

M_V = m_v - 5 log d + 5 = 2.5 - 5 log (40) + 5

M_V = 2.5 - 8.5 + 5 = -1

M_{bol} = M_V + B.C. = -1 + (-1.7) = -2.7

Log (L*/L) = 0.4 (M_{bol} - M_{bol}*) = 0.4(4.63 - (-2.7)) =

Log (L*/L) = 0.4(7.33) = 2.93

2 Log (R*/R) = Log (L*/L) - 4 Log (T_{eff}*/T_{eff})

Using the table of Bolometric corrections vs. T_{eff} we find for B.C. = -1.7:

Log (T_{eff}) = 3.52

And we know Log (T) = Log (5760) = 3.76

So: 2 Log (R*/R) = 2.93 - 4(3.52 - 3.76) = 2.93 + 0.96

= 3.89

Log (R^*/R) = 3.89/ 2 = 1.946

and: R^*/R = 88.3 (Since antilog (1.946) = 88.3)

Now:

(a) The apparent discrepancy is the difference in the results:

% diff. = [(109.4 R = 88.3R)/ 109.4 R] x 100% = 19% approx.

(b) Both values suffer from about the same degree of uncertainty, so that a mean value, e.g.:

R_{av} = ½(109. 4 + 88.3)R

would be the best.

c) The interferometry method has a significant probable error owing to the fact that Beta Pegasi is at *the very limit of practical application for the trigonometric parallax distance technique* ($\approx$ 50 pc). A similar uncertainty also appears in the photometric method, e.g. in the determination of M_V and hence also M_{bol}, log (L^*/L) and R^*/R. Thus, there is no appreciable benefit in selecting one method over the other.

In this case, the simpler (non-interferometer) method is the more logical choice.

3) The most expeditious way to solve for this set is to

first prepare a calculation table based on the key logs for the assorted relationships. These are: log (M*/M), log (L*/L), 4 Log (T_{eff}^*/T_{eff}) and 2 log (R*/R). The method of obtaining each of the preceding is simply based on computations we've already shown, given the data indicated, e.g. (B - V), M_V.

STAR Name	log (L*/L)	log (M*/M)	4 log(T*e/Te)	2 log(R*/R)
Rigil Kent.	0.208	0.0594	-0.12	0.328
Spica	4.172	1.192	2.56	1.612
Fomalhaut	1.252	0.358	0.76	0.492
Suhail	4.252	— *	-0.96	5.212
RESULTS TABLE	MASS	Luminosity	RADIUS	Temperature (eff.)
Rigil Kent.	1.15 M	1.6L	1.46R	5370 K
Spica	15.6M	14 860L	6.4R	25 120 K
Fomalhaut	2.28M	18 L	1.7R	9120 K
Suhail	? *	17 865 L	403R	3300 K
TABULATED LOGS & RESULTS for PROBLEM No. 3			* Non Main Sequence	

The upper part of the Table presented shows the assorted log computation results. Note that Suhail doesn't have a proper mass value, since the mass-luminosity index isn't valid, being off the main sequence.

The lower region of the table shows the assorted mass, luminosity, radius values based on the preceding log table. (E.g. we work out the radius, using the given log results for temperature and luminosity, such as demonstrated in Problems 1, 2. Again, Suhail's mass is in question. Note, we can still obtain a radius and luminosity for it by relying exclusively on the (B - V) and M_V values and making reference to an H-R (Hertzprung -Russell) diagram to get estimates.

(b) Surface temperatures can be obtained using the Bolometric corrections table, and referencing the (B - V) to the appropriate log (T_{eff}).

Thus, for Rigil Kent, (B - V) = 0.71 which corresponds to log (T_{eff}) = 3.73 from the Table. Then taking the antilog, we obtain T_{eff} = 5370 K. In the case of Spica, (B - V) = -0.23 and we see tabulated values given in the table for -0.20 (4.32) and -0.25 (log T_{eff} = 4.46), so we interpolate the value for -0.23 (e.g. 3/5 of the scale between -0.20 and -0.25) to get log T_{eff} = 4.40 or T_{eff} = 25 120 approx.

Extra Problem:

For each of the stars listed in the Table on the opposite page (175), plot the position on the H-R diagram shown on p. 372. For convenience, you can make a photocopy of the H-R diagram then plot the 4 stars listed on it.

Bear in mind what each axis of the diagram shows and the information and–or corrections to the listed data needed to complete it.

Are all of the stars on the Main sequence? If not, name the exceptions.

XV: *Physical Aspects of the Stars*

What exactly is a star? Basically it can be thought of as an immense, controlled fusion bomb, bound by gravity. It manufactures heavier elements through the nuclear fusion process and generates heat and light as a result.

I leave the physics details of nuclear fusion for the Appendix, but in this chapter we will treat the gross stellar properties arising therefrom. With regard to this, it's a of interest to note we have more confidence regarding the nature of the physical interior of stars and the Sun than the interiors of planets like the Earth. The reason is that the physical laws to be applied are simpler once one is dealing with an ionized gas or plasma.

It is the nature of the ionized gases and plasma that mean we can't treat stars in the usual sense of a pure gas – say based on the ideal gas laws. The Sun, for example, has a relative density that's 1.4 times that of water (1000 kg m^{-3}) while the common terrestrial gases with which we're familiar have much lower densities.

If a star isn't an ordinary ball of gas, then what is it? A clue is provided by the characteristically high temperatures of stellar interiors. These temperatures are so high, e.g. 10 million degrees Celsius, that no normal gas can exist. Thus it's nearly totally ionized allowing a much greater compression of stellar material without significant deviation from the perfect gas law of physics – though that law must now be modified. The modified form is written:

$P = (N_o + N_e) kT$

Where N_o is the number of atoms of all kinds, both neutral (unionized) and ionized and N_e is the number of free electrons (the greater this number, the higher the degree of ionization). The total gas pressure therefore is the sum of the two pressures:

$P = N_o kT + N_e kT$

The second term is known as the 'electron pressure". Then the ratio between the gas and electron pressure is:

$P/P_e = 1 + N_o / N_e$

An alternative way of expressing the stellar gas equation of state is:

$P = (\rho / \mu H) kT = \rho RT / \mu$

Where ρ is the gas density and μ is the mean molecular weight, with H the mass of the hydrogen nucleus or about 1.7×10^{-27} kg)

An expression for μ in terms of the chemical composition of a star may be arrived at from a simple tabulation involving the number of atoms of the relevant elemental species and the numbers of free electrons arising from their respective ionizations.

We begin by noting that for any star we must have: $X + Y + Z = 1$

Where X denotes the fraction of hydrogen, Y the fraction of helium, and Z the fraction of all the heavier elements. For our tabulation we set out:

//////////	**HYDROGEN**	**HELIUM**	**HEAVIER**
No. of atoms	$X\rho / H$	$Y\rho / H$	$Z\rho / H$
No. of electrons	$X\rho / H$	$2Y\rho / H$	$\tfrac{1}{2}AZ\rho / H$

Therefore we have:

N (total) = $(2X + 3Y/4 + \tfrac{1}{2}Z\rho / H)$

And finally, after simplifying:

$1/\mu = (2X + 3Y/4 + \tfrac{1}{2}Z)$

For most stars to which we will apply the above equation of state, Z represents only a negligible fraction (e.g. in the Sun about 0.2% or 0.002)so we can therefore simplify further:

Thus, let $Z = 1 - X - Y$ so:

N(total) = $\rho / H (2X + 3Y/4 + (1 - X - Y)/2)$

N(total) = $\rho / H (3X/2 + Y/4 + \tfrac{1}{2})$

$= \rho / 4H (6X + Y + 2)$ or:
$\mu = 4 / (6X + Y + 2)$

This is a good approximation except in the cool, outermost layers of a star. Then since Z is negligible we may replace Y with $(1 - X)$ and obtain:

$$\mu = 4/(3 + 5X)$$

Besides gas pressure, there is also the radiation pressure:

$$P_R = aT^4/3$$

where a is the radiation density constant which is equal to: 7.55×10^{-16} Jm^{-3} K^{-4} and T is the absolute temperature (in K degrees). The equation above is valid only if the material is in thermodynamic equilibrium (i.e. the distribution of radiation is that of a black body). This is generally true, but some exceptions do exist.

In dealing with stars, energy considerations also arise. In particular, the thermal and gravitational energy of a star are very closely related. In a perfect gas, the total thermal energy is found by multiplying the number of particles N by the degrees of freedom, f, possessed by each particle. The thermal energy per unit volume is then: ½Nf kT.

The number of degrees of freedom f is also related to the ratio of specific heats (γ) of the material by:

$$\gamma = (f + 2)/f$$

where γ is the ratio of the specific heat at constant pressure to the specific heat at constant volume, or:

$\gamma = C_p / C_v$

Using the equation for a perfect gas and introducing the thermal energy per kg, U, we find:

$U = P / (\gamma - 1)\rho$

Bear in mind the material inside a star is highly ionized and in physics a fully ionized gas or plasma is mono-atomic, for which $\gamma = 5/3$.

Using the previous equation for U in conjunction with the virial theorem, it can be shown that for a star with $\gamma = 5/3$:

$2U + \Omega = 0$

where U is the thermal energy for the whole star and Ω is the gravitational energy such that:

$\Omega = - G \int_o^R M(r)\, dM(r) / r = - GM^2 / 2R$

But the potential energy, $V = 2\Omega = - GM^2 / R$

This can be put into an even more useful form based on the kinetic theory of gases, for which:

$P = \rho v^2 / 3$

And: $- \Omega = 3 \int P dV = 3 \int (P/\rho)\, dm$

$3 \int (P/\rho)\, dm = 3 \int v^2 dm / 3 = m v^2 = 2K$

Where K denotes the (gravitational) kinetic energy, which is also expressed:

$K = GM^2/R$

Then in terms of the gravitational energy, Ω:

$-\Omega = 2K$ or $2K + \Omega = 0$

But: $\Omega = -GM^2/2R$ so: $K = -\Omega/2 = GM^2/R$

Which checks out. We are left with these conclusions:

1) $2K + \Omega = 0$ applies to any spherical system in equilibrium where K is the gas kinetic energy and also the gravitational kinetic energy. ($K = 3/2(\gamma - 1)U$). As we saw before:

$U = P/(\gamma - 1)\rho$ so: $P/\rho = U(\gamma - 1)$

Then: $3 \int (P/\rho) \, dm = 3 \int U (5/3 - 1) \, dm = 2K$

2) The binding or total energy of a star E_T is then:

$E_T = K + V = GM^2/2R + (-GM^2/R) = -GM^2/2R$

Or: $E_T = \Omega/2 = -K$

Thus, the total energy of a star is negative and equal to half the gravitational potential energy or the negative of the gas kinetic (or gravitational kinetic).

3) We conclude from the preceding that if for some reason E_T decreases, then K increases but Ω decreases (e.g. the sphere must contract).

Practice Problems:

1) For a uniform sphere with a polytropic index n = 0 for uniform density, show that V = -6/5 (GM²/R). Take the potential to be:

$\Omega = 3/(n-5)(GM^2/R)$

For any polytropic gas sphere.

Solution:

   Polytropic gas spheres are basically mathematical entities used for modeling of actual stars. As usual, some basic assumptions are made (often in terms of temperatures, pressures, potential energies etc.) and these are then used to develop one or more "polytropic" models to test to see if they can work for a given star. Or, more likely, be employed as a guide to model a star.

   A primary objective is to develop a basis for a self-gravitating sphere. In the most desirable of cases, one works to attain a simple relationship between the pressure P, and density (ρ) of a form:

$\boldsymbol{P} = K(\rho)^{(1 + 1/n)}$

where K and n are constants, and n is known as "the *polytropic index*" and K the "polytropic constant".

The polytropic index n can be defined:

n = 1/ (γ - 1)

where 'γ' is the ratio of specific heats.

The problem thus asks the reader to regard the hypothetical uniform sphere as having a polytropic index n = 0 associated with a sphere of uniform density, and adopting:

Ω = 3/ (n − 5) (GM²/ R)

Then we may simplify using n = 0 to get:

Ω = 3/ (0 − 5) (GM²/ R) = -3/5 (GM²/ R)

We have already shown that: V = 2Ω so:

V = 2Ω = 2 [-3/5 (GM²/ R)] = -6/5 (GM²/ R)

2) In a given layer for a 10 solar mass star, the composition is 90% hydrogen, 10% helium. Assume the layer is in local thermodynamic equilibrium. The temperature is taken to be 3.14×10^6 K. Take the density of the material as ρ = 20.2 kg m⁻³. Find the relative contributions of gas and radiation pressure in the layer and the thermal energy per kilogram of the material. (Assume the gas is totally ionized).

Solution:

Begin by finding the mean molecular weight of the material given that: X = 0.9 so we use:

$\mu = 4/(3 + 5X) = 4/(3 + 5(0.9)) = 0.533$

The gas pressure, $P = (\rho/\mu H)kT = \rho RT/\mu$

So P =

$(20.2 \text{ kg m}^{-3})(8.3 \times 10^3 \text{ JK}^{-1})(3.14 \times 10^6 \text{ K})/0.533$

$P = 9.88 \times 10^{11} \text{ Nm}^{-2}$

The radiation pressure is: $P_R = aT^4/3$

Where $a = 7.55 \times 10^{-16} \text{ Jm}^{-3}\text{K}^{-4}$

So that $P_R = 1/3 (7.55 \times 10^{-16} \text{ Jm}^{-3}\text{K}^{-4})(3.14 \times 10^6 \text{ K})^4$

And: $P_R = 2.46 \times 10^{10} \text{ Nm}^{-2}$

The total pressure is:

$P_T = \rho RT/\mu + aT^4/3 =$

$9.88 \times 10^{11} \text{ Nm}^{-2} + 2.46 \times 10^{10} \text{ Nm}^{-2} \approx 10^{12} \text{ Nm}^{-2}$

Thus, the overall contribution of radiation pressure is only on the order of 0.02 or about 2% of the total pressure.

The thermal energy per kg is: $U = P/(\gamma - 1)\rho$

Or: $U = (10^{12} \text{ Nm}^{-2})(5/3 - 1)(20.2 \text{ kg m}^{-3})$

$U = 7.4 \times 10^{10} \text{ J kg}^{-1}$

To get some physical idea of the situation treated in problem (2), the reader may refer to the diagram below showing the inner onion-like structure of a spherical gas system such as found in a star:

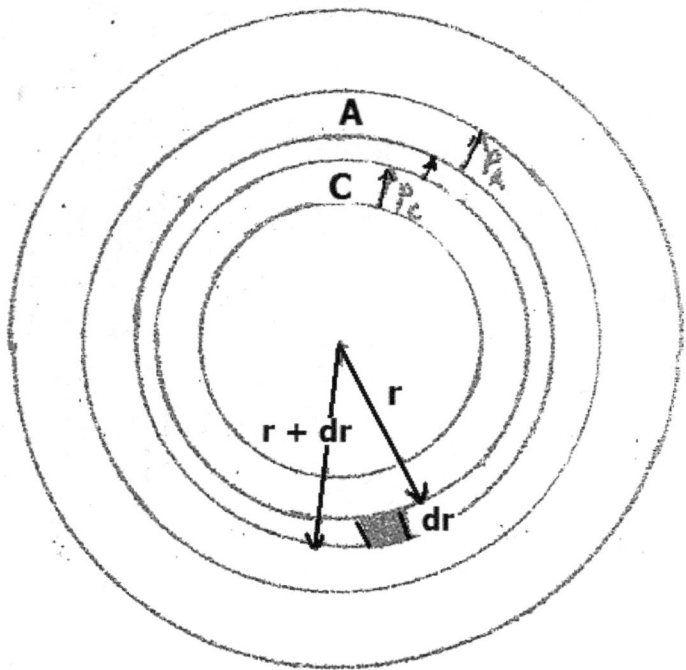

Fig. 1: Basic Stellar structure referred to Prob. 2

Here, three layers, A, B and C are depicted in the interior of the ten solar mass star. The layer B was the one under consideration in the problem identified by the increment or differential radius, dr. Thus, since layer A is further removed from the core, it would be at LTE but at a cooler temperature, while C would be at LTE at a hotter temperature. (Thus a **gradient** of temperature, $\Delta T/ \Delta R$, exists from the core to the surface).

From the gas equations of state we can easily see that greater (combined) gas pressures accompanies greater stellar depths. Note therefore from the diagram that the respective layer pressures are in the relation:

$P_C > P_B > P_A$

And in all cases the pressure characteristic of the layer is directed outwards, i.e. toward the surface. Clearly then, some force must be acting in the opposite direction to resist the outward pressure and maintain the star in a state of balance. Otherwise the star would simply inflate, or contract.

The fact stable stars like the Sun exist, shows that a pressure-gravity balance must obtain. To see how this can be quantified, consider the element of layer B shown of width dr and lying between r and (r + dr). Let P be the pressure at r and let the increment in P be dP. The difference in pressure dP represents a force, -dP acting on the mass element considered in the direction of increasing r.

The mass of the element considered is: ρ dr. Then the force of attraction between M(r) e.g. the mass enclosed inside the sphere of radius, r and ρ dr is the same as that between a mass M(r) at the center and ρ dr at r. By Newton's law this attractive force is given by:

$F = G\, M(r)\, \rho\, dr\, /\, r^2$

Since the attraction due to the material outside r is zero, we should have for equilibrium:
$-dP = G\, M(r)\, \rho\, dr\, /\, r^2$

Or

$$dP/dr = -G\,M(r)\,\rho\,dr/r^2$$

Note that P has been used to denote the total pressure. Thus P is the sum of the gas pressure and the radiation pressure.

Consider now the mass of the shell between layer A and layer C. This is approximately, $4\pi\,r^2\,\rho\,dr$, provided that dr is small. The mass of the layer is the difference between $M(r + dr)$ and $M(r)$ which for a thin shell is:

$$M(r + dr) - M(r) = (dM/dr)\,dr$$

Equating the two expressions for the mass of the spherical shell we obtain:

$$dM/dr = 4\pi\,r^2\,\rho$$

The two equations, for dP/dr and dM/dr represent the basic equations of stellar structure, without which the innards of a star would be inaccessible to investigation.

A third equation of stellar structure may be derived using by using the equation for dM/dr in combination with the fact that a star's luminosity is produced through the consumption of its own mass. This may be expressed mathematically as:

$$dL/dM = \varepsilon$$

where ε denotes the rate of energy generation. For the proton-proton cycle (for stars like the Sun- and *designed for cgs units!*):

$\varepsilon = 2.5 \times 10^6 \, (\rho \, X^2) \cdot (10^6/T)^{2/3} \exp[-33.8(10^6/T)^{1/3}]$

Worked Problem (3):

a) Derive a third equation for stellar structure where dL/dr is the subject. (Hint: Make use of the energy generation form for dL/dM).

b) From one or more equations of stellar structure, obtain an estimate for the Sun's central temperature and pressure. (Take the solar radius R = 7 x 10⁸ m and the solar mass M = 2 x 10³⁰ kg, and the density ρ = 1400 kgm⁻³.)

Solution:

a) We use: $dL/dM = \varepsilon$ and $dM/dr = 4\pi r^2 \rho$

Then by *the chain law for derivatives*:

$(dL/dM)(dM/dr) = dL/dr = \varepsilon \, (4\pi r^2 \rho)$

b) We approximate: $dP/dr = - G M(r) \rho \, dr/ r^2$
To: $P/R = G M \rho / r^2$

$P = (6.7 \times 10^{-11} \, Nm^2 kg^{-2})(2 \times 10^{30} \, kg)(1400 \, kgm^{-3})/ R$

Where R = 7 x 10⁸ m

Then: $P = 2.6 \times 10^{14}$ Pa

For the estimate of the central temperature we use:

$T = \mu P / \rho R$

Where we know $\mu = 0.57$ for a fully ionized H-plasma.

Then: T =

$(0.57)(2.6 \times 10^{14}$ Pa$)/(1400$ kgm$^{-3})(8.3 \times 10^3$ JK$^{-1})$

$T = 1.2 \times 10^7$ K

Other Problems:

1) Using the formula for the molecular weight, show that for electrons only: $\mu = 2 / (1 + X)$.

2) In the atmosphere of a star at a particular layer, the temperature is 5650 K and there are approximately 1.45×10^{19} free electrons. If the total gas pressure is 8.3×10^3 Pa find: a) the electron pressure, and b) the total number of neutral and ionized atoms.

3) Use the temperature obtained in the last sample problem above to obtain an estimate of the rate of energy generation in the solar core. Assume an 80% hydrogen content in the core. Also use the approximate equation for sample worked problem (3) to obtain an estimate of the solar luminosity. By what amount is this in error if the actual luminosity is $L = 3.9 \times 10^{26}$ W.

4) A star of 10 solar masses is being investigated. According to a stellar model used for the star: X = 0.9, Y = 0.1 at a point r/R* = 0.5. The temperature characterizing the layer which contains the point is given as: T(r) = 4.7 x 10^6 K and the density ρ(r) = 10 kgm^{-3} . Would such a model be correct and consistent? Show using the information in this chapter. (Assume the point is midway in a shell dr = 0.002R* and R* = 6R$_{(solar)}$.)

Project:

Consider a model of a star for which: M/M$_s$ = 0.6, L/L$_s$ = 0.57 and R/R$_s$ = 0.64, where the denominator subscript refers to solar values. The table below contains the run of variables in a portion of the stellar model over a radius differential Δr = 0.16R:

r/R	m(r)/M	L(r)/L	logp(r)	logT(r)	logρ(r)	X(r)
0.58	0.802	1.000	15.183	6.494	0.537	16.7
0.62	0.843	1.000	15.015	6.451	0.412	20.5
0.66	0.877	1.000	14.841	6.393	0.297	20.9
0.70	0.908	1.000	14.655	6.318	0.185	32.2
0.74	0.934	1.000	14.451	6.237	0.063	41.3

Using the method of *finite differences*, determine whether *this portion of the star* is totally radiative, totally convective, or partially radiative and partially convective. If the last alternative is the case, then locate the critical radius (r/R) at which convection sets in. Assume the plasma is totally ionized so the ratio of specific heats is γ = 5/3.

XVI. Stellar Evolution Basics

In the cores of stars, for billions of years, heavier elements have been produced by the nuclear fusion of lighter elements, such as hydrogen and helium into much heavier ones. Very massive stars ultimately became unstable, their cores imploding even as their outer regions explode into space as supernovae. In this violent process, all the newly synthesized heavier elements are ejected into space.

These dispersed elements then become the 'birth material' for newer, heavier generations of stars, not to mention planets, moons, asteroids and other objects. Elemental "Evolution", then, is an ongoing process where chemicals are built up inside stars and later disseminated.

In the Sun, for example, two distinct nuclear fusion processes occur: 1) the proton-proton cycle, and 2) the carbon-nitrogen cycle. In the first of these (the easier one because it has fewer reactions):

$H_1 + H_1 + e^- \rightarrow D_2 + neutrino + 1.44\ MeV$

$D_2 + H_1 \rightarrow He_3 + gamma\ ray + 5.49\ MeV$

$He_3 + He_3 \rightarrow He_4 + H_1 + H_1 + 12.85\ MeV$

The top line shows two protons fusing to yield deuterium (*heavy hydrogen*) with a positron and neutrino emitted, along with 1.44 MeV of energy.[1]

---

[1] This is a unit of energy with which most readers may not be familiar. One electron volt (eV) is defined as the energy needed to push a single electron charge of 1.6×10^{-19} Coulomb through 1

Empirical evidence of this reaction is obtained from gallium detectors, of the neutrinos given off, which are within 1-2% of what theoretical models predict. In the second fusion reaction, the deuterium combines with a proton to give the isotope helium 3, along with a gamma ray and 5.49 MeV energy. In the final fusion, two helium-3 nuclei combine to yield one helium-4 nucleus, along with two protons, and 12. 85 MeV energy. Note that the two protons commence the cycle anew, so that the generation of nuclear energy is ongoing.

This means that the central core of the Sun will become heavier and heavier, as more and more helium is produced. This despite the fact that the Sun as a whole is losing an amount of mass of roughly 4×10^6 metric tones per second. To see how this works, recall that the sum of fractional elemental abundances for the Sun or any star:

$$X + Y + Z = 1$$

where X denotes hydrogen fraction, Y is the helium fraction, and Z the heavier elements' fraction. After assignment of relative atomic weights, electrons etc. one obtains:

$$1/\mu = 2X + 3Y/4 + Z/2$$

where μ is the mean molecular weight of the relevant solar gas. Now, since Z in the Sun is negligible (Z less than 0.02)we can write:

---

volt. This would be: 1.6×10^{-19} Joule of energy. One MeV is one million times that: $(10^6) \times 1.6 \times 10^{-19}$ Joule.

$$\mu = 4/(8X + 3Y)$$

A simple check using the above can be made for an estimated current fractional abundance in the core, say X = 0.73, Y = 0.25, to some future abundance, X = 0.70, Y = 0.28 (Z fraction remaining essentially constant). It will be found that the internal core mean molecular weight increases from m = 0.61 to 0.62. Hence, molecular weight of the core plasma increases as the hydrogen fusion reactions proceed.

At some stage, when nearly the entire core is helium a new "burning" phase will be ushered in (at higher temperature), such that the following reaction series, known as the 'triple alpha' process, kicks in:

He 4 + He 4 → Be 8 + gamma (- 95 keV)

Be8 + He 4 → C 12 + gamma + 7.4 MeV

Here, the two alpha particles (helium nuclei) first fuse to give unstable beryllium and a gamma ray (gamma), with 95 keV energy *absorbed*. Then the beryllium fuses with a helium-4 to give carbon−12 plus a gamma ray and 7.4 MeV energy given off. In this way a new cycle commences, leading to a heavier molecular weight core. Each successive burning phase is less efficient than its predecessor, as can be seen by comparing the energy given off in the triple alpha process to that given off in the proton-proton cycle.

The key thing to bear in mind in terms of a stable phase (i.e. 'Main sequence') star like the Sun is that it is in pressure-gravity balance. The outer gas pressure balances the weight of its overlying layers. Any condition likely to disrupt this balance is therefore of paramount interest.

The stable lifetime of the Sun depends on how long before it consumes ninety percent of the hydrogen in its core. Theoretical investigations using data from nuclear reaction rates and cross sections suggest the Sun's Main Sequence lifetime at 8-10 billion years. Since it already has spent 4.5 billion of those years, there are anywhere from 3.5 to 5.5 billion years remaining.

How do we quantify the time for stars like the Sun to remain in a stable, say *pressure –gravity* equilibrium? (I.e. such that, $dP/dr = - G M(r) \rho \, dr/ r^2$ from the previous chapter). The tool we use to do this is known is an "H-R diagram" for **Hertzsprung-Russell diagram**, after the two astronomers that first devised it.

In such diagrams there are two "quality dimensions" displayed on the vertical axis and on the horizontal. In the former we have the "**Absolute magnitude**" which represents a measure of absolute brightness, denoted M, and also doubles as an index for what we call *Luminosity* (right scale). In terms of absolute magnitude the lower the value, the greater the brightness. (As we saw already in Chapter III.)

An illustrative H-R diagram is shown in Fig. 1.

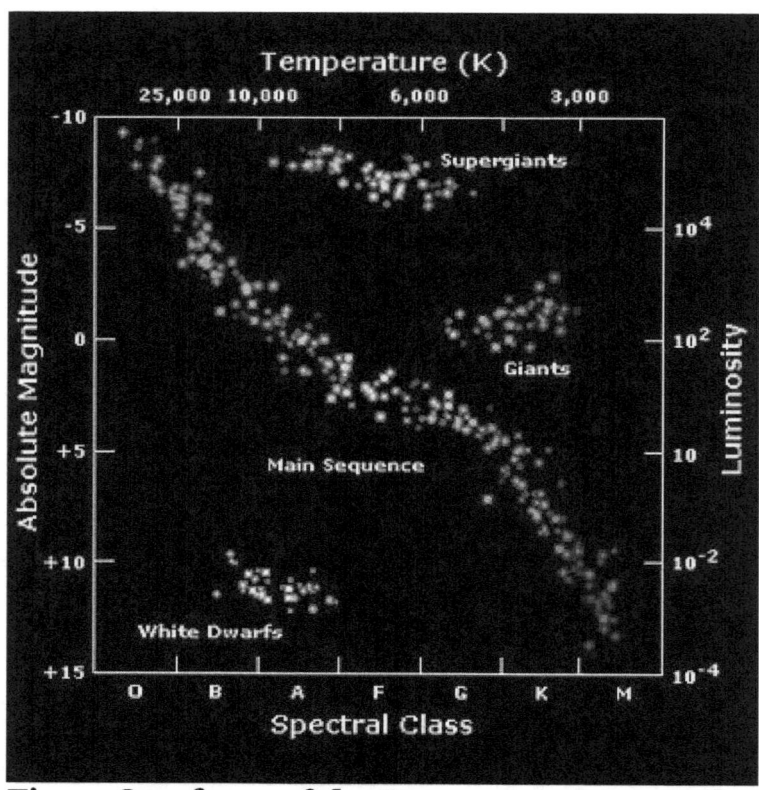

Fig. 1: One form of the Hertzsprung-Russell Diagram

Note again that the scale is also logarithmic, so that a star of (-10) say will always be 100 times brighter than a star of (-5), so five magnitudes difference in M corresponds to 100 times brightness ratio. The other quality dimension is for the horizontal axis which is usually associated with the temperature or **spectral class**. In terms of this lettering, class O stars are always the hottest, and class M the coolest. Thus, we have also can double the spectral class as a temperature scale (top of diagram). Within this representation framework all the key domains - star colors, star ages, star distances, star energy use, can be integrated and depicted in their relations if one can interpret the quality axis properly and use them in specific ways.

How does distance enter? Well, the absolute magnitude is defined in terms of the brightness – magnitude a given star has at a distance of 10 parsecs, or 32.6 light years. Thus, the M-value allows comparison of stellar brightness at a standard distance.

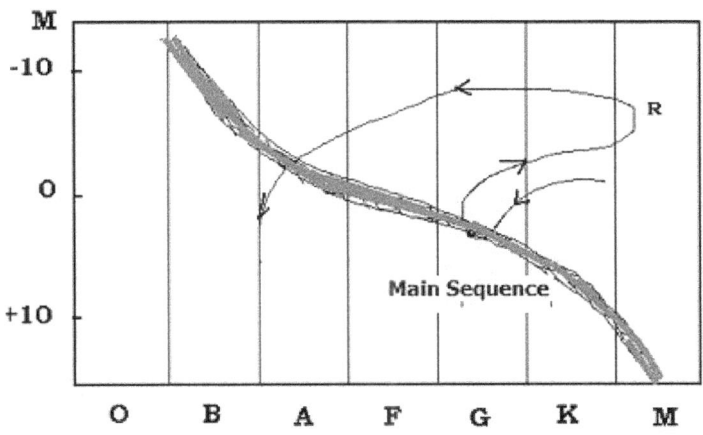

Fig.2: H-R diagram with solar evolutionary track.

Stellar ages in the H-R diagram enter by way of the position for the star's evolutionary track (sketched above in Fig. 2 for the Sun, as a wavy thin line) makes with a band called "the Main Sequence" with the latter being the superposed domain for all stable stars.

When only ten percent hydrogen remains (X= 0.10,

Y = 0.87 for fractional abundances, say), the Sun is no longer able to generate sufficient energy from its core nuclear reactions to balance the weight of overlying layers. According to a well-known physical principle (the virial theorem- see previous chapter), the Sun's core must contract. The contraction converts gravitational potential energy into thermal (heat) energy that heats the core. By now, hydrogen burning has moved to a peripheral shell around the core, and is ignited by the core heating process. The ignition creates radiation pressure that forces the outer shells, layers to expand. This same radiation, however, is now emitted from a much larger surface area. The result of this combination of circumstances is that the Sun becomes a Red Giant.

The details and theoretical consistency of such diagrams are generally checked by plotting brightness (on the stellar magnitude scale) and spectral index or color index (B-V) for a variety of open star clusters, such as the Pleiades.

Going back to Fig. 2, the dot on the Main Sequence indicates roughly where the Sun is currently in its evolution. As can be seen, its path to the Red Giant (R)region remains ahead of it, as does its subsequent collapse to a compressed white dwarf star, with the track veering down and to the left.

The time on the main sequence can easily be estimated using:

$\log (T_{ms}) = 10.11 + \log (M/M_s) - \log (L/L_s)$

The preceding equation is based on the assumption

that the main sequence lifetime concludes when only about 10% of hydrogen remains. At this point the star begins evolution away from the Main Sequence.

Example Problem: Find the lifetime on the Main sequence for Sirius, with $M/M_s = 2$, $L/L_s = 11$.

Solution:

$\log(T_{ms}) = 10.11 + \log(2) - \log(11)$

$\log(T_{ms}) = 10.11 + 0.3 - 1.04 = 9.37$

And antilog$(9.37) = 2.35 \times 10^9$

Hence, Sirius' Main sequence lifetime is 2.35×10^9 yrs.

In the case of a star much more massive than the Sun pre-Main Sequence collapse occurs within an interstellar gas and dust cloud, but the Main Sequence is joined at a higher position, corresponding to greater luminosity. Thereby, astronomers have determined the empirical relation for mass and luminosity.

That is, the luminosity (L) is proportional to the mass (M)raised to the 3.5 power. What this also means is that the massive star initiates its Main Sequence lifetime at higher temperatures, including higher core temperatures. And like a fast-living human, its energy is consumed much more rapidly, so the course and duration of evolution is dramatically speeded up. Fusion reactions are much more diverse than in the Sun, obviously because a greater range of heavier elements is produced.

A key transition point occurs after carbon is formed in the core, and reaches a critical density and temperature to detonate. The resulting *deflagration* includes the core separating from exploding outer layers, turning the star into an instant nuclear factory. Nickel and iron are formed as well as lighter elements in the shells including: magnesium, sulfur, silicon, manganese, chromium and a host of lesser atomic weight elements such as B, F, C, Ne, O, N.

If the star survives carbon detonation, its end is still heralded by formation of nickel-iron in the core. The nuclear reactions become endothermic, absorbing energy instead of generating it. This means the star's radiation pressure and gas pressure supporting the outer layers is radically decreased. Collapse of the layers occurs, with oxygen ignited in one of them precipitating the spectacular explosion we call a supernova. The most recent significant such event was designated 1987A, in The Large Magellanic Cloud. Gamma ray line radiation from the decay of Cobalt 56 (Co 56) was detected in this event[2] indicating earlier formation of an unstable Nickel (Ni 56) core.

In such cosmic cataclysms the star's outer layers explode into space, while its core collapses to form a neutron star or black hole. It is the outer layers - containing magnesium, silicon, sulfur and especially carbon, expelled into space, that sets the stage for the future evolution of life on other worlds.

---

[2] Arnett, D. and Bazan, G.: *Nucleosynthesis in Stars: Some Recent Developments*, in **Science**, Vol. 276, 30 May, 1997, p. 1359.

What is the importance of stellar evolution then? Just this: that biological life forms cannot arise unless the fundamental element for life - carbon, has been first manufactured in stars. It also shows that the stars themselves evolve *at different rates*, the more massive stars at faster rates than the less massive (like the Sun). The role of massive stars, therefore, is to expedite elemental production and chemical evolution by making available more complex elements which become the building blocks for planets as well as life, on a more rapid time scale than would otherwise be possible.

A multitude of embryonic stars can be detected right now with infrared telescopes, as well as the Hubble Space Telescope, collapsing out of interstellar dust and gas analogous to phases 1 and 2 of the evolutionary track in the H-R diagram for the Sun. Some of these *T Tauri* stars are in the well-known Orion Nebula, about 1600 light years away. On an H-R diagram like that shown for the Sun, these embryo stars would be on the track approaching the Main Sequence.

The T Tauri and other observations clearly demonstrate that the creationist hypothesis can't be valid. If it were, no new stars should be forming!

Observational H-R Diagrams:

These are used to plot the positions of actual stars, often which occur in star clusters, and using the absolute visual magnitude (M_v) for the vertical axis against the (B- V) color index for the horizontal which takes the place of the spectral class. (See Chapter XII) Such an H-R diagram is shown below.

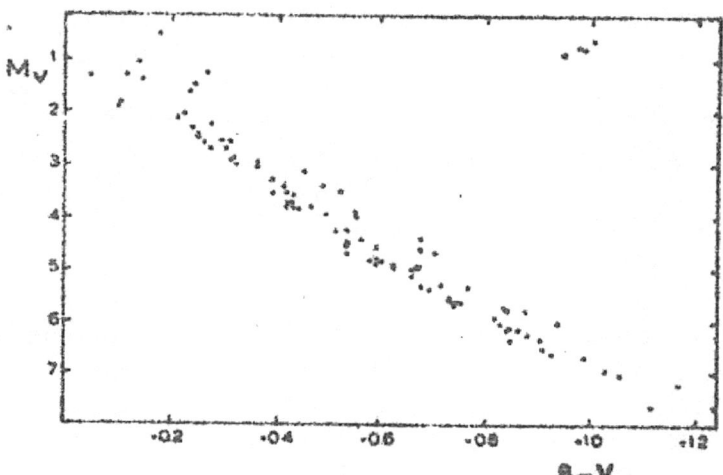

Fig. 3: Observational H-R diagram for an open cluster

One of the first things that must be done to compare the above to a theoretical H-R diagram is to convert the absolute visual magnitude M_v to an absolute bolometric magnitude M^*_{bol} using a bolometric correction. Thus we would find: $M^*_{bol} = M_v + B.C.$ and:

$\log (L/L_s) = 0.4(M_{bol} - M^*_{bol})$

where M^*_{bol} denotes the star's bolometric magnitude.

Project:

For the open star cluster NGC 2632, below:

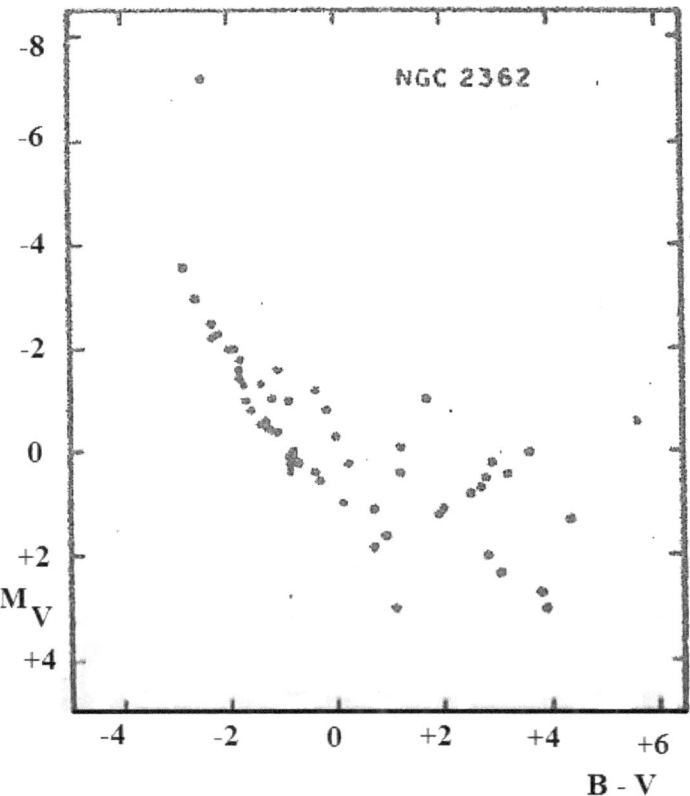

Construct a *theoretical* H-R diagram. For the vertical axis you will need log (L/L$_s$) or simply L/L$_s$. For the horizontal axis you will display the surface temperature, T instead of (B − V). After completion, lightly shade in the main sequence for this star cluster. (Hint: You can make use of the bolometric corrections table on p. 140.)

XVII. Stellar Atmospheres & Radiative Transfer

Modeling stellar atmospheres, just as computing self-consistent stellar models, is similarly a very complex undertaking that often requires we make basic assumptions. However, it is possible to approach the topic at a rudimentary (but still useful level) by appealing to simplified assumptions and not being so rigorous as to expectations.

The *"gray atmosphere"* is one such simplifying assumption. First, we need to present some preliminaries.

The Planck function describes the distribution of radiation for a black body, and can be expressed:

$B(\lambda) = \{(2\ hc^2)/\lambda^5\}\ [1/\exp(hc/\lambda kT) - 1)]$

where h is Planck's constant, c is the speed of light, T is the absolute temperature, k is the Boltzmann constant, and λ, the wavelength. In the plane-parallel treatment, we take layers of the gases in a stellar atmosphere to be like layers of a "sandwich", where ds is an element of length or path perpendicular to the layers.

```
--------------------
--------------------
------------------]  -------------------------------------- ds
--------------------
```

This as opposed to employing curved layers (as would technically be the case), for which the math is many times more complex! A more detailed plane-parallel model atmosphere is depicted below, in Fig. 1:

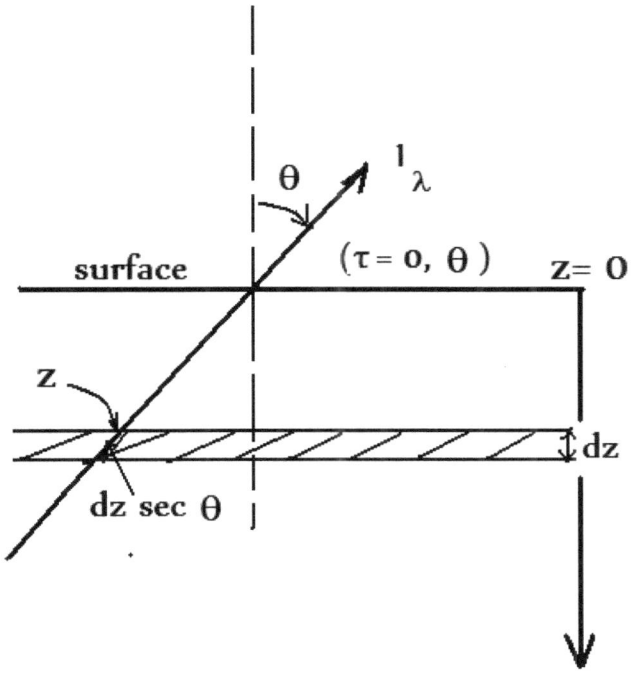

Fig. 1: Plane –Parallel model stellar atmosphere

As a beam of radiation (I_λ) passes through parallel-layered stellar gases, there will be *emission and absorption* along the way. The "**source function**" specifies the ratio of one to the other and can be expressed:

$S(\lambda) = \varepsilon(\lambda) / \kappa(\lambda)$

where λ again denotes wavelength, ε(λ) is the emission coefficient, and κ(λ) the absorption coefficient.

In the case of simple radiation transfer in a static model stellar atmosphere (e.g. nothing changes with time), we have the relation of radiation intensity I(λ) to source function S(λ):

dI(λ)/ds = -κ(λ) I(λ) + κ(λ) S(λ)

= κ(λ) [S(λ) − I(λ)] - 0

or I(λ) = S(λ)

Now, for *a black body*, I(λ) equals the Planck function B(λ) :

So, in effect, we have:

S(λ) = I(λ) = B(λ)

And this is a condition which implies LOCAL THERMODYNAMIC EQUILIBRIUM or *LTE*

   LTE does NOT mean complete thermodynamic equilibrium!(E.g. since in the outer layers of a star there is always large energy loss from the stellar surface)

   Thus, one only assumes the emission of the radiation is the same as for a gas in thermodynamic equilibrium at a temperature (T) corresponding to the temperature of the layer under consideration.

Another way to say this is that if LTE holds, the photons always emerge at all wavelengths.

Now, in the above treatment, note that the absorption coefficient was always written as: κ(λ) to emphasize its wavelength (λ) dependence.

However, there are certain specific treatments for which we may eliminate the wavelength dependence on absorption, and simply write e.g. k – to denote having the same absorption value at ALL wavelengths!

This is what is meant by the "*gray atmosphere*" approximation.

Here is a specific application of the gray atmosphere approximation. In a particular integral, let the surface flux

π(F_o) = 2 π (I(cos (θ)) = π [a(λ) + 2(b(λ)/3]

and $F_{λo}$ = S(λ) τ(λ) = 2/3

which states that the flux coming *out of the stellar surface* is equal to the source function at the *optical depth* τ = *2/3*. This is the very important '*Eddington-Barbier' relation* that facilitates an understanding of how stellar spectra are formed.

Once one then assumes LTE, one can further assume κ(λ) is independent of λ (gray atmosphere) so that:

$\kappa(\lambda) = \kappa$; $\tau(\lambda) = \tau$ and $F_{\lambda o} = B_\lambda(T(\tau = 2/3))$

Thus, the energy distribution of F_λ is that of a black body corresponding to the temperature at an optical depth $\tau = 2/3$.

From this, along with some simple substitutions and integrations (hint: look at the Stefan-Boltzmann law!) the interested reader can easily determine:

$\pi(F_o) = \sigma(T_{eff})^4$ and $T_{eff} = T(\tau = 2/3)$

where σ is the Stefan-Boltzmann constant. Thus, the temperature at optical depth 2/3 must equal the *effective temperature*!

Intensity and Moments of Intensity:

For more detailed analyses, more complex mathematics is required. However, what follows should not be beyond the ability of anyone who's done first year calculus. What we want is to first be able to write or express the emergent intensity as it's depicted in Fig. 1. This is done by first writing:

$I_\lambda(0,\theta) = \int_0^z B_\lambda(\tau) \exp[(-\tau_\lambda / \cos\theta)] d\tau / \cos\theta$

where $B_\lambda(\tau)$ is the Planck function, and the integral is taken from z = 0 to the point z in the interior.

Next, since we're dealing with the passage of radiation *out of the star*, we need the relevant equation of transfer. This is best dealt with using

integral with dx, and hence recasting the emergent intensity in the form:

$$I_\lambda(0,\theta) = \int_0^\infty B(x)\, e^{(-x/\cos\theta)}\, dx/\cos\theta$$

Then the appropriate equation of transfer would be:

$$dI = j\, dx - \sigma I(\theta)\, dx = (j - \sigma I(\theta))\, dx$$

or: $dI/dx = j - \sigma I(\theta))$

Or, after some further manipulation, and replacing x with τ:

$$(\cos\theta)\, dI/d\tau = I(\theta) - j/\sigma$$

This is the important equation, in terms of emergent intensity, that embodies the conservation of radiant energy (i.e. no more radiation can flow out of a star's surface than can be generated within it and which approaches that surface).

Next, we want to be able to obtain an even more improved basis for our calculations and this entails getting the moments of the intensity. These are defined as follows:

$$J = 1/4\pi \int_{4\pi} I(\theta)\, d\omega \quad \textit{(Mean Intensity)}$$

Where $d\omega$ is an element of solid angle - defined as (A/r^2) for a sphere, for example. Thus a sphere with surface area $A = 4\pi r^2$ has solid angle $(4\pi r^2 / r^2) = 4\pi$ steradians. If we are only dealing with a sliver of emergent beam of area $0.01\pi\ r^2$ then the element of solid angle is:

$$d\omega = (0.01\pi\ r^2 / r^2) = 0.01\pi\ \text{sr}$$

Or 1/400 the volume of a sphere.

The next moment is:

$$H = 1/4\pi \int_{4\pi} I(\theta) \cos\theta\ d\omega$$

This is defined as the *"net flux"* or the net energy breaching the stellar surface in units of net energy per second per unit area of that surface.

Finally, we come to the last moment of intensity:

$$K = 1/4\pi \int_{4\pi} I(\theta) \cos^2\theta\ d\omega$$

This is the *energy density*.

At this stage, we are in a position to use each of the above moments to further manipulate the equation of transfer. We start by multiplying the original equation of transfer: $(\cos\theta)\ dI/d\tau = I(\theta) - j/\sigma$ through by $1/4\pi \int_{4\pi} I(\theta)\ d\omega$ to get:

$$dH/d\tau = 1/4\pi \int_{4\pi} I(\theta)\ d\omega - 1/4\pi \int_{4\pi} j/\sigma\ d\omega$$

This makes use of the definition: $dH/d\tau = J - j/\sigma$

Problem: Verify the above definition using the definitions of J, H and the transfer equation.

Next, we multiply the equation of transfer through by H, or $1/4\pi \int_{4\pi} I(\theta) \cos\theta \, d\omega$ to get: $dK/d\tau = H$ or more simply: $4\pi H = $ const. So, $dH/d\tau = J - j/\sigma = 0$, and the equation of transfer now becomes:

$$(\cos\theta) \, dI/d\tau = -I(\theta) + 1/4\pi \int_{4\pi} I(\theta) \, d\omega = -I + J$$

From here a number of specific assumptions are made in order to not have to evaluate the integral. The main one is the Eddington approximation which will apply to the quantities J, H and K. We will also discriminate the radiation intensity I into two components: I_1 (in the forward direction) and I_2 (in the backward direction). We can then write as follows:

1) $J = \frac{1}{2}(I_1 + I_2)$

Problem: Prove the above by integrating I in the forward and backward directions.

Hint: Use $1/4\pi \int_{4\pi} I(\theta) \, d\omega =$

$1/4\pi \int_{0}^{2\pi} \int_{0}^{\pi/2} I_1 \sin\theta \, d\theta \, d\phi + 1/4\pi \int_{0}^{2\pi} \int_{0}^{-\pi/2} I_2 \sin\theta \, d\theta \, d\phi$

Similarly, we have:

2) $H = \frac{1}{4}(I_1 - I_2)$

3) $K = J/3$

Now, we focus on the boundary in Fig. 1, and note that here the optical depth $\tau = 0$, and we must have $I_2 = 0$ also. Since $I_2 = 0$ then: $H = \frac{1}{4} I_1$ and $J = \frac{1}{2} I_1$ so clearly: $J = 2H$.

Further, $K = J/3$ or $(\frac{1}{2}I_1)/3$ so $K = H\tau + \text{const.}$

This follows, since we had: $dK/dC = H$ or $dK = H\, d\tau$ and we know $\tau = 0$, hence $H\tau + \text{const.}$ on integration. From this it follows that:

$J = H(2 + 3\tau)$ and $K = J/3 = 2H/3$

At the boundary everywhere.

And since $H = \frac{1}{4}(I_1 - I_2) = \text{const.}$

$I_1(\tau) = H(4 + 3\tau)$ and $I_1(\tau) = 3H\tau$

A special case occurs if the mean intensity $J = B$, the Planck function, then (since $B \approx \sigma T^4/\pi$):

$J = H(2 + 3\tau) = \sigma T^4/\pi$

Therefore, the boundary temperature (T_o) approaches the value of the effective (or surface) temperature when $\tau = 0$. So we have the basic relationship:

$\sigma T_o^4/\pi = 2H$

And: $\sigma T^4 = \sigma T_o^4/2\ [\alpha + 3\tau]$

In the limit of this approximation, $T_{eff}^4 = 2 T_o^4$
And hence:

$T_{eff} = (2)^{1/4} T_o = 1.189 T_o$

Sample Problem (1)

1) a) Estimate the specific intensity $I (\theta=\pi/4)$ if the surface flux from the Sun is 6.3×10^7 Jm^{-2} s^{-1}.

(b) Find the mean intensity if $I (\theta=\pi/4)$ is taken over all space.

(c) Find the effective temperature of the Sun and the boundary temperature (T_o) and account for any difference.

(d) Estimate the net flux, H, passing through the Sun's surface.

Solution:

The specific intensity is defined from:

$\pi(F_o) = 2 \pi (I(\cos (\theta))$

And for $\theta = =\pi/4$, then $\cos (\pi/4) = \sqrt{2}/2$ and:

$I = \pi(F_o) / 2 \pi (\sqrt{2}/2) = \pi(F_o) / \pi (\sqrt{2})$

Therefore:

$I = (6.3 \times 10^7$ Jm^{-2} $s^{-1}) / \pi(\sqrt{2})$

$I \approx 1.4 \times 10^7$ Jm^{-2} s^{-1}

(b) If I is taken over all space (i.e. 4π steradians), the mean intensity is:

$$J = 1/4\pi \int_{4\pi} I \, d\omega = 1/4\pi \, (1.4 \times 10^7 \, Jm^{-2} \, s^{-1}) \, 4\pi$$

So: $J \approx 1.4 \times 10^7 \, Jm^{-2} \, s^{-1}$

(c) The effective temperature is obtained using:

$\pi(F_o) = \sigma(T_{eff})^4$

So: $T_{eff} = [\pi(F_o) / \sigma]^{1/4}$

Where $\sigma = 5.67 \times 10^{-8} \, W \, m^{-2} \, K^{-4}$

Is the Stefan-Boltzmann constant. Then:

$T_{eff} = [6.3 \times 10^7 \, Jm^{-2} \, s^{-1}/ 5.67 \times 10^{-8} \, W \, m^{-2} \, K^{-4}]^{1/4}$

$T_{eff} \approx 5800 \, K$

The boundary temperature is found from:

$T_{eff} = (2)^{1/4} \, T_o = 1.189 \, T_o$

Or: $T_o = T_{eff} /1.189 = 5800K/ 1.189 \approx 4800 \, K$

    The boundary temperature differs because of being referenced to a different optical depth. The boundary temperature (T_o) *approaches* the value of the effective (or surface) temperature when $\tau = 0$, but this still exhibits a difference in layers so will not be exactly the same!

(d) The net flux (H) passing through the Sun's surface is estimated using: $\sigma T_o^4 / \pi = 2H$.

Therefore: $H = \sigma T_o^4 / 2\pi$

Where: $\sigma T_o^4 = (5.67 \times 10^{-8}\ W\ m^{-2}\ K^{-4})(4800K)^4$

$\sigma T_o^4 = 3.01 \times 10^7\ W\ m^{-2}$

$H = (3.01 \times 10^7\ W\ m^{-2}) / 2\pi = 4.79 \times 10^6\ W\ m^{-2}$

Sample Problem (2):

Consider the solar half-sphere and the energy going into it each second. We know on average photons are absorbed after traveling a distance with optical depth $\tau = 1$ in the propagation direction. Averaged over all directions this corresponds to a vertical optical depth of $\tau = 2/3$. Based on this find:

a) The energy going into the half sphere each second

b) The change in (a) over each absorption and re-emission over vertical optical depth.

c) The total absorption and total emission and the relationship between then over all space.

d) Find each of the above for the star *ξ Ophiuchi* which has a (B – V) color index of -0.30.

Solution:

 a) This is just: $\sigma(T_{eff})^4 = \pi F = \Delta 2\pi S = \Delta 2\pi I$

b) Over a vertical optical depth one has $\tau = 2/3$, then: $\Delta 2\pi S/\Delta \tau = \pi F/(2/3) = 3\pi F/2$ or: $\Delta S/\Delta \tau = 3\pi F/2(2\pi) = 3F/4$

c) Over all space, the total emission = $4\pi S$ and the total absorption = $4\pi J$, and by radiative equilibrium: $4\pi S = 4\pi J$ so that $S = J$.

d) We need to use the Table on page 140 to identify the $(B - V)$ index with the log of effective temperature, log T_{eff}, for which we find: log T_{eff} = 4.65 because we see: $(B - V) = -0.30$ corresponds to log T_{eff} = 4.65.

antilog (log T_{eff}) = antilog (4.65) = 44,700

So: T_{eff} = 44,700 K

Then: $\pi F = \sigma(T_{eff})^4 =$

$(5.67 \times 10^{-8}$ W m^{-2} K$^{-4})(4.47 \times 10^4$ K$)^4 =$

$(5.67 \times 10^{-8}$ W m^{-2} K$^{-4})(4 \times 10^{18}$ K$^4) = 2.2 \times 10^{11}$ W m^{-2}

Whence:

$\Delta S/\Delta \tau = 3\pi F/4\pi = 3(2.2 \times 10^{11}$ W m$^{-2})/4\pi$

Or: $\Delta S/\Delta \tau = 5.4 \times 10^{10}$ W m^{-2}

Also: $2\pi S = \pi F$ or:

$S = \pi F/(2\pi) = (2.2 \times 10^{11}$ W m$^{-2})/2\pi$

Or: $S = J = 3.5 \times 10^{10}$ W m^{-2}

Sample Problem (3):

A spherical star of radius R emits radiation of uniform intensity I in all directions. Determine the mean intensity and the flux of the radiation at a distance r from the star.

Solution:

The mean intensity and the net flux have already been defined as the first two moments of intensity:

$$J = 1/4\pi \int_{4\pi} I(\theta) \, d\omega$$

And

$$H = 1/4\pi \int_{4\pi} I(\theta) \cos\theta \, d\omega$$

The above integrals are each over the solid angle subtended by the star. Since the star radiates in all directions then I = const. for every direction at which r comes from the star and is zero for all other directions.

Let the element of solid angle $d\omega$ be:

$$d\omega = \sin\theta \, d\theta \, d\phi$$

where θ denotes the angular radius subtended by the star. Then:

$$J(r) = I/4\pi \int_0^{2\pi} d\phi \int_0^{\pi/2} \sin\theta \, d\theta = \tfrac{1}{2} I(1 - \cos\theta)$$

And:

$$H(r) = 1/4\pi \int_0^{2\pi} d\phi \int_0^{\pi/2} \cos\theta \sin\theta \, d\theta = \pi I \sin^2\theta / 4\pi$$

Then: $H(r) = I \sin^2\theta / 4$

In the limit of small θ the angular radius satisfies:

$\sin\theta = R/r$, then use the trig identity:

$\sin^2\alpha = \frac{1}{2}(1 - \cos 2\alpha)$

in $J(r) = \frac{1}{2} I(1 - \cos\theta)$

So: $J(r) = I \sin^2(\theta/2) = I(R/2r)^2 = IR^2/4r^2$

Similarly, $H(r) = I(R/r)^2/4 = IR^2/4r^2$

So, in the limit as $r \ggg R$, $J(r) \approx H(r)$

Other Problems:

1) A star has a gray atmosphere for which the Eddington approximation: $T^4 = \frac{3}{4} T_e^4 (\tau + 2/3)$ is valid, where T_e denotes the effective temperature. Use this approximation to obtain the fraction of outward intensity escaping from the star's surface.

2) The star Suhail has a $(B - V)$ color index of $+1.7$. Use this and the Table on p. 140 to obtain the net flux (H) passing through the Suhail's surface. How might you estimate the intensity I from this and the mean intensity J?

3) For many stars, the solar constant S can be computed if its angular diameter is known. If the angular radius of a star is: $\alpha = (R/r)$ with r the distance to Earth and R the star's linear radius then: $\pi F = S (r/R)^2$

(Note: that α is measured in *radians*)

a) If the Sun's angular radius is 959.63 arcsec then find the solar constant S. (Hint: you can use the solar flux πF already computed from sample problem 1).

b) Find the solar constant S for α Lyrae (Vega) if we know (from Hanbury and Brown's measurements) that its angular diameter is 0.0032 arcsec, and it has a (B- V) color index of 0.00. Show all working and state any assumptions.

Project:

The fraction of specific intensity I originating above some specified optical depth $\tau 1$ can be computed from:

$I(\tau < \tau 1)/I = 1 - (0.6\tau 1 + 1) \exp[-\tau 1]$

Use this relation to prepare a table ranging from:

0.0 $< \tau 1 <$ 10.0

to show the fractions of emitted intensity from differing optical depths. Use this information to answer these questions: i) How much emitted intensity originates below $\tau 1 = 5$? Ii) How much originates above $\tau 1 = 2$? How much below $\tau 1 = 1$

XVIII. Solar Corona Physics

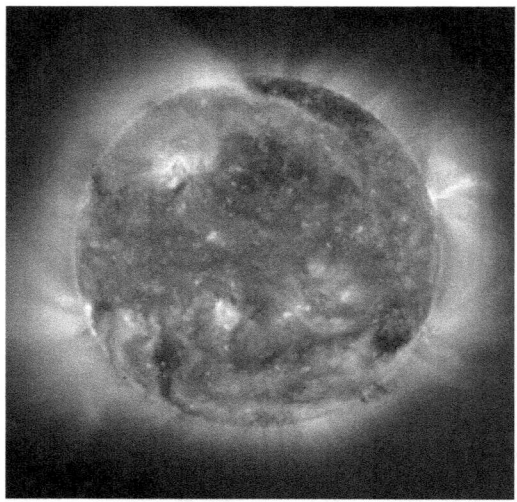

Image of coronal holes taken by Yohkoh.

As the new solar cycle ramps up much attention again will focus on the Sun's corona. Thanks to new imaging tools, with increased optical as well as temporal resolution, we ought to be getting unparalleled images of the outermost atmosphere of the nearest star. Of particular relevance here will be the Atmospheric Imaging Assembly (AIA), which was built for the *Solar Dynamics Observatory* by the Lockheed Martin Solar Astrophysics Laboratory.

Since being installed, the AIA's been taking many high resolution images of the corona in multiple wavelengths, from the near to the extreme ultra-violet (UV). Shown in the graphic is one of the representative images taken at 211Å, taken last October. This corresponds to the corona's characteristic temperature of 2×10^6 K. It is also the wavelength most sensitive to magnetically active regions (ARs) and we see white lines superimposed on those field lines.

Among the interesting questions pursued in the past century was whether the corona was static or not. In a static case its boundary would be more or less fixed, there'd be no expansion even in times of high solar activity.

A static corona superficially appears to be quite reasonable but that's why we need to test this is so. The first one to do this was Sydney Chapman. He began by first assuming the condition for hydrostatic equilibrium applied:

$dp/dr = -\rho \{GM_s/r^2\}$

where G is the usual Newtonian gravitational constant, and rho defines *the plasma density* for the corona, while M_s is the mass of the Sun, and r the distance from the solar center:

$\rho = n(m_p)$

with n the number density for protons

The coronal pressure (P) is given by:

$P = 2nT$

Provided both protons and electrons are assumed to have the same temperature.

The *thermal conductivity* of the corona is dominated by electron thermal conductivity and takes the form:

$$\kappa = \kappa_0 \, T^{5/2}$$

for typical coronal conditions the value of κ is about 20 times the value of copper at room temperature.

Now, the *coronal heat flux density* is:

$$q = -\kappa \nabla T$$

A static corona means *heat inputs cancel heat outputs* so that the divergence:

$$\nabla \cdot q = 0$$

Assuming a spherical symmetry for the corona one can write:

$$1/r^2 \, [d/dr \, (r^2 \, \kappa_0 \, T^{5/2} \, dT/dr)] = 0$$

Obviously the preceding assumptions mean there must be some distance where the coronal temperature becomes zero.

From the above equation one should be able to show:

$$d(T^{7/2}) = 7/2 \, (F \, T_0^{5/2}) / \, 4 \, \pi \, \kappa_0 \, d(1/r) = C \, d(1/r)$$

where C is a constant.

The integral is:

$$T_0^{7/2} - T^{7/2} = C[\, 1/R_0 - 1/r \,]$$

Now, set the temperature at infinity (T) to zero and obtain:

$C = R_o T_o^{7/2}$

which fixes the total flux at:

$F = 2/7 \ [4 \pi R_o \kappa_o T_o]$

After another step, one finds:

$T(r) = T_o (R_o / r)^{2/7}$

this gives the temperature T at a distance from the Sun = r. This is based on using a defined value (say $T_o = 2 \times 10^6$ K) at a defined distance, say $R_o = 7 \times 10^8$ m.

For example, at the Earth's distance ($r = 1.5 \times 10^{11}$ m) one would find: $T = 4.3 \times 10^5$ K

This seems fine, until one examines the pressure.

Analogous to the temperature formalism, we have, the pressure p(r) at some distance r defined by:

$p(r) =$

$p(R_o) \exp [7/5 \ GM_s \ m_p / 2 \ T(R_o) \ R_o \{(R_o / r)^{5/7} - 1\}]$

Now, if one allows r to approach infinity, e.g. $r \to \infty$ an interesting thing occurs in the equation, as we can see. That is, the denominator of the first term in the end brackets becomes so large (R_o / ∞) that the first term vanishes.

Then we are left with the expression for the pressure:

$$p(\infty) = p(R_o) [\exp - 7k/5 * 1/ T(R_o) R_o]$$

where 'k' denotes a constant composed of all the constant quantities in the previous eqn. (G, M, m_p etc)

Substituting the given values into the above, one finds $p(R_o)$ multiplied by a factor

$$\exp[0] = 1$$

The reason is that the exponential of a very small and negative valued magnitude $\to 0$

Then:

$$p(\infty) \approx p(R_o)$$

But **this can't be** since the pressure of the coronal base would then be *the same as the value at infinity!*

This led astrophysicists to conclude an unphysical result, and that **the static coronal model couldn't be accurate**.

If the static model were accurate, the pressure at infinity should be zero, $p(\infty) = 0$, not a small finite pressure that's effectively equal to the coronal base pressure. This finding led to the further investigations that disclosed a solar "wind" had to flow outwards from the corona.

Sample Problem (1):

Find the value of the constant C at the solar corona boundary.

Solution:

We use: $C = R_o T_o^{7/2}$

So: $C = (7 \times 10^8 \text{ m})(2 \times 10^6 \text{ K})^{7/2} = 7.9 \times 10^{30} \text{ m K}^{7/2}$

Sample Problem (2):

The thermal conductivity of the corona is given approximately by the empirical formula:

$\kappa = 1.8 \times 10^{-10} (T^{5/2} / \ln \Lambda) \text{ W m}^{-1} \text{ K}^{-1}$

Where $\ln \Lambda \approx 20$.

From this estimate the mean flux at the coronal boundary (R_o) and at a distance $2 R_o$ where the mean temperature is 10^6 K. Compare the value of the first with the total luminosity of the Sun ($L = 3.9 \times 10^{26}$ W)

Solution:

First, we note that the coronal conductivity in the empirical formula has the same form as: $\kappa = \kappa_o T^{5/2}$

If we separate out the $T^{5/2}$ factor, then it follows that:

$\kappa_o = [(1.8 \times 10^{-10})/ \ln \Lambda] \text{ W m}^{-1}$

Or: $\kappa_o = (1.8 \times 10^{-10})/ 20$ W m^{-1} = 9×10^{-12} W m^{-1}

The flux at the coronal boundary (lower layer) is defined:

$F = 2/7 \ [4 \pi R_o \kappa_o T_o]$

Where $T_o = 2 \times 10^6$ K at a distance $R_o = 7 \times 10^8$ m.

Then: F =

$2/7 \ [4 \pi (7 \times 10^8 \text{ m})(9 \times 10^{-12} \text{ W m}^{-1} \text{ K}^{-1})(2 \times 10^6 \text{ K})]$

Or: F ≈ 4.5×10^4 W

The total solar luminosity is: L = 3.9×10^{26} W

Or, more than: $(3.9 \times 10^{26}$ W$)/ 4.5 \times 10^4$ W

= 8.6×10^{21} times greater.

At a distance r = 2 R_o the temperature is given by:

$T(r) = T_o (R_o / r)^{2/7} = (2 \times 10^6 \text{ K})(R_o /2 R_o)^{2/7}$

$T(r) = (2 \times 10^6 \text{ K})(1/2)^{2/7} = 1.6 \times 10^6$ K

To get the flux, we assume: $F = 2/7 \ [4 \pi R \kappa T]$

Where: $\kappa = 1.8 \times 10^{-10} (T^{5/2} / \ln \Lambda)$ W m^{-1} K^{-1}

= $1.8 \times 10^{-10} [(1.6 \times 10^6 \text{ K})^{5/2} / 20]$ W m^{-1} K^{-1}

= 3.1×10^4 W m^{-1} K^{-1}

Therefore: F =

2/7 [4π (14×10^8 m) (1.1×10^3 W m^{-1} K^{-1}) 1.6×10^6 K]

F = 2.5×10^{20} W

Other Problems:

1) It is said that the Earth is "*enveloped in high temperature coronal material*". Verify this by obtaining the temperature of coronal material at the Earth's distance from the Sun: d = 1.5×10^{11} m.

2) In the textbook *Stellar Atmospheres* (1978), author Dimitri Mihalas claims (p. 523) that: "*At 10^6 K the conductivity of coronal plasma far exceeds ordinary laboratory conductors.*" Verify that this is so by obtaining the conductivity *for that temperature* and comparing it to the following electrical conductivities of lab materials:

Silver: 6.3×10^7 S/m

Copper: 5.9×10^7 S/m

(Where the units S/m denotes **Siemens** per meter.)

At what distance from the coronal boundary would this temperature occur?

3) Find the coronal pressure at the Earth's distance if that pressure as a function of distance r is given by:

p(r) = p_o exp{- $7R_o$[1 - (R_o /r)$^{5/7}$]/ 5H}

where H ≈ 10^8 m is known as the "scale height". (Take p_o = 0.2 dynes/cm² so you can use c.g.s. units to solve the problem.

4) According to Mihalas (*op. cit.*, p. 525) the coronal "flow velocity" v can be deduced from the flux F based on the equation:

F = $4 \pi r^2$ n (v)

where n denotes the particle density, i.e. per unit volume. Compute this velocity: a) at the distance of the Earth (See: #1) and b) at the distance associated with the coronal temperature from Problem #2.

You may use the equation (Mihalas, p. 524):

n(r) = n_o $(r/R_o)^{2/7}$ exp{- $7R_o$[1 – $(R_o/r)^{5/7}$]/ 5H}

taking n_o = 4 x 10^8 /cm³

(Hint: Convert n_o to the corresponding value per m³)

5) Compare the velocity obtained above to the critical velocity for the corona defined as equal to the ion sound speed, and given by (Mihalas, p. 526):

V(r_c) = $[2kT_o/m]^{1/2}$

Where k is Boltzmann's constant and m is the ion mass = 1.7 x 10^{-24} kg.

6) Which of the preceding problem results disclose that the corona is a *dynamic entity* and not in any equilibrium?

XIX. Galaxies & Density Waves

We now return to astrophysics, namely pertaining to the spiral galaxies, and what engenders the spiral shape. Of course, one can easily duplicate the shape using something as simple as dropping some cream in a cup of tea. One simply stirs up the black tea with a spoon, then drops several droplets of cream into it, and 'Voila!' a mini-spiral "galaxy" forms before one's eyes. Of course, things aren't quite so simple for real galaxies!

First, it's well to bear in mind that the particular density model (for a given galaxy) will vary depending on the conditions for a given galaxy. In principle then, there can be differing density wave models, and these pertain to a number of variables, factors such as: how tightly wound the spiral is (there are different grades which are assigned, e.g. Sa, Sb, Sc etc.), the degree of axial symmetry of the galaxy, and the modeling assumptions - in particular the potential-gravitational fields (V(r)) imposed on the system which determine the locations of orbital resonance in conjunction with the equations used.

The gravitational potential energy is defined according to:

$V(r) = - GMm/r$

where G is the Newtonian gravitational constant, m is the unit or elemental mass within the galaxy, M is associated with the central mass concentration, and r is the distance from m to M. The negative sign, as usual, indicates a *bound system*.

Interestingly, the use of density wave model development is largely contingent on the Boltzmann equation which is also used in space plasma physics. Thus:

$$\partial f/\partial t + v \text{ grad } f + F/m\, \partial f/\partial t = (\partial f/\partial t)_c$$

where the $\partial f/\partial t$ are partial derivatives, and $(\partial f/\partial t)_c$ is the time rate of change in f (the **velocity distribution function**) due to collisions, i.e. between masses within the system. Technically, the Boltzmann eqn. is applied to FLUIDS and for that purpose the galaxies to which density wave approaches or models are applied are modeled firstly in the fluid format. (It is easier when dealing with an agglomeration of some 100 or 200 billion separate stars and associated orbits to think of them as comprising a "fluid" as opposed to say, 100 billion separate bodies to be treated in a 100 billion -body problem of celestial mechanics!)

The referencing of stars, their locations and movements meanwhile embodies particulate approaches that are more kinematical in nature (but often less amenable to consistency with the density wave approach). Orbital assumptions, declarations are not simple by any means, and merely because a source says or asserts that *"The stars in the inner part of a galaxy move faster than the density wave/s and the stars in the outer parts move slower than the wave/s."* should not be taken too literally without posing a lot of further questions. (And one could argue here that "apples" and "oranges" are being compared because the two entities, stars and density waves arise from differing backgrounds - kinematic-particle based and fluid mechanical, wave based.)

For example, what class is the spiral? How tightly wound? One must recognize too that an orbit in a spiral galaxy that *appears* closed (e.g. elliptic) in one reference frame may not be so in another. As an example, assume the (polar) coordinates for a galactic rotating frame are given as (r, φ) with:

$d\varphi/dt = d\Theta/dt - \Omega_p$

where Ω_p is the angular velocity of the rotating frame. Then orbits are described by a Hamiltonian (recall the Hamiltonian adds kinetic and potential energies of the system):

$H = \frac{1}{2}(p_r^2 + p_\varphi^2/r^2) + V(r) - p_\varphi \Omega_p$

where the p_r, p_φ are the particle momenta referred to the associated coordinates, and V(r) is the gravitational potential. The point is that H can change *depending on the coordinates*, and what is presented for the previous frame as $H = E - J \Omega_p$ (with simplification, $p_\varphi = J$) may well be different for another frame.

Second, we see from this that the question as to why the spiral pattern is not affected by stars much further out cannot really be properly answered unless a full vetting of the assumed density waves for the particular galaxy is presented. In this sense, one recognizes that a full analysis of density waves for a galaxy - call it "Barred G1"- is needed before one can say stars in a given G1 region (e.g. inner or outer) "move faster or more slowly" than the *waves* at that place. We need to know then: the physical conditions for the establishment of the density waves at location r1 in G1 and r10 in G1 where the r's denote radial distances from the center with r10 = 10 (r1).

Another problem in dealing with density wave models is the fact they are mainly based on the *mode* chosen for particular dynamical wave equations that can be applied to the fluid framework. (Note: A "mode" is a standing wave that can be supported by a disk of given dimensions, mass.) More broadly, most astronomers who work in this narrow specialist area use the term interchangeably with **Fourier m-component**. (And it should be understood here that one of the main tools is Fourier analysis of the waves, but alas Fourier analysis is only taught usually to those who take advanced Calculus or analysis courses).

As an example, a particular Fourier coefficient, call it a_n, applicable to a wave - may be defined:

$$a_n = 1/\pi \int_{-\pi}^{\pi} f(x) \cos mx \, dx$$

where m is the Fourier m-component.

What types of modes can one have in these models? One is the "global" or m= 1 mode. Then there are the unstable (m= 2) modes.

Whether one mode or another appears (or is used in a spiral galaxy modeling) is critical since it may well determine at what stage a barred spiral develops, if at all. Alas, another complexity enters here since mode analysis is not simply a stand alone but also incorporates a subtle aspect called "marginal stability analysis" wherein one will solve for a quantity Q and if it is very close to 1 one has the case of marginal stability and tightly wound modes or in the case of spirals, around the Sa class. The trouble is that when one seriously incorporates any heating of the disk for whatever reason (say a massive central black hole sucking up matter and generating much radiation) then the desired values of Q are soon out of range, making it impossible for a given spiral structure to sustain itself.

Lastly, whenever one considers density waves in galaxies, it's important to bear in mind there remain enormous stumbling blocks even when applied to the simplest models of galactic disks (e.g. "zero thickness" disks). One of these arises from potential theory. Thus, the perturbed gravitational field at one location depends on the density perturbation at *every other location*. How will you know, *ab initio*, that the density perturbation at location r, φ, z say, does not accelerate the associated wave (in the fluid rest frame) to a higher velocity than any stars at the same or near location? You don't unless you investigate! What does it mean to "investigate"? It means a full bore mathematical modeling procedure to locate where all the *Lindblad orbital resonances* are, since these can speed up the star. (We also want to know where the Landau damping regions are, which can impose a retardation of the waves.)

In short, what I've shown is that density wave analysis as applied to galaxies is a field almost to itself in terms of being amenable to general understanding. And complexity is often compounded because the mere posing of a question to do with a particular spiral galaxy's form almost always introduces a number of tacit assumptions that may not be applicable at all. This, of course, is the difficulty when dealing in generalities, as opposed to specific cases, examples.

Too bad we can't have a model based on as simple a fluid description as what happens to cream dropped into a stirred cup of tea!

Sample Problem (1):

Assume that the Sun moves in a nearly circular orbit about the galactic center with the radius of orbit R = 10^4 parsecs. It then makes an entire orbital revolution in 2×10^8 years.

a) Using Kepler's third law obtain an estimate for the mass of the Milky Way galaxy. (Recall there are 2×10^5 AU per parsec.

b) Estimate the gravitational potential at the Sun's location if its mass m = 2×10^{30} kg (take 3.26 Ly = 1 parsec and 9.46×10^{15} m = 1 Ly.)

Solution:

(a) By Kepler's 3rd law: $(M+ m)(P1/ P2)^2 = k(a1/ a2)^3$

Or in simplified form, with a2 = 1 AU, P2 = 1 yr:

$(M+ m)(P_1)^2 = (a_1)^3$

Where: $a_1 = a = (10^4 \text{ pc})(2 \times 10^5 \text{ AU/pc}) = 2 \times 10^9 \text{ AU}$

And $P_1 = P = 2 \times 10^8$ yrs

Solution: Since the mass of the Sun, $m << M$ (the mass of the galaxy), we may write:

$(M)(P)^2 = (a)^3$

Or: $M = (a)^3 / (P)^2$

$M = (2 \times 10^9 \text{ AU})^3 / (2 \times 10^8 \text{ yrs})^2$

$M = (8 \times 10^{27}) / (4 \times 10^{16}) = 2 \times 10^{11}$ (m) the mass of the Sun.

(b) Since $m = 2 \times 10^{30}$ kg, then $M \approx 2 \times 10^{11}$ (m)

$M \approx 2 \times 10^{11} (2 \times 10^{30} \text{ kg}) \approx 4 \times 10^{41}$ kg

The gravitational potential at the Sun's distance is:

$V(r) = -GMm/r$

Or:

$V(r) =$

$-(6.7 \times 10^{-11} \text{ N-m}^2/\text{kg}^2)(4 \times 10^{41} \text{ kg})(2 \times 10^{30} \text{ kg}) / r$

Where: $r = (10^4 \text{ pc})(3.26 \text{ Ly/pc})(9.46 \times 10^{15} \text{ m/Ly})$

Or: $r = 3.1 \times 10^{20}$ m

Then: $V(r) = -5.3 \times 10^{61}$ N-m² / 3.1×10^{20} m
$V(r) \approx -1.7 \times 10^{41}$ J

Sample Problem (2):

The number χ of epicycle oscillations per orbit about a galactic center, is given by the ratio of the star's epicycle frequency (κ_o) to its orbital angular speed, Ω.

Or: $\chi = \kappa_o / \Omega$

a) Compute Ω for the Sun using the information in sample problem (1) and find the Sun's epicycle frequency (κ_o) if we know $\chi = 1.35$. (Note: if the ratio χ is integral, or *a non-decimal number*, we say the orbit is "*closed*". If non-integral it is not closed). Because of differential rotation rates and the local angular velocity (Ω_L) differing from the inertial value, the solar region is open. See the open model below:

b) Compute the local angular velocity (Ω_L) for the Sun if: $m(\Omega - \Omega_L) = n \kappa_o$ where $m = 2$ and $n = 1$

Solution:

(a) In sample problem (1) we saw that the Sun's period around the galactic center is: 2×10^8 yrs. We need to obtain this in seconds:

$T(s) = (2 \times 10^8 \text{ yrs.})(365.25 \text{ days/yr.})(86,400 \text{ s/day})$

$T(s) = 6.3 \times 10^{15}$ s

Then: $\Omega = (2\pi \text{ rad})/T(s) = 2\pi \text{ rad}/6.3 \times 10^{15}$ s

$\Omega \approx 10^{-15}$ rad s^{-1}

We have: $\chi = \kappa_o / \Omega = 1.35$ and thus:

$\kappa_o = 1.35 \ \Omega = 1.35 \ (10^{-15} \text{ rad s}^{-1}) = 1.35 \times 10^{-15}$ rad s^{-1}

(b) Re-arrange the given expression:

$m(\Omega - \Omega_L) = n \kappa_o$

By deliberately choosing to have the Sun complete n orbits in the rotating frame of reference (while executing m epicycle oscillations).

Then we can write:

$\Omega_L = \Omega - n\kappa_o/m = \Omega - \kappa_o/2$

Therefore:

$\Omega_L = 10^{-15}$ rad s^{-1} $- \frac{1}{2}(1.35 \times 10^{-15}$ rad s$^{-1}) =$

3.2×10^{-16} rad s^{-1}

Other Problems:

1) The specific angular momentum for a star located in a galactic disk is: $p_\varphi = J = r^2 \Omega(r)$.

In the rotating galactic frame the key ratio $\chi = \kappa_0 / \Omega$ becomes: $2(\Omega - \Omega_p) / \kappa_0$.

Where Ω_p is the angular velocity of the rotating frame.

a) Obtain a value of J for the Sun, using the solar data from the previous (worked) problems. Then use the same to obtain an estimate for Ω_p.

b) Finally, obtain a value for the rotating frame Hamiltonian: $H = E - J\Omega_p$

Where E is the energy of the solar orbit in the galaxy. (Hint: $E \approx -GMm/r$, where m is the solar mass, and M the mass of the galaxy).

3) There are 3 conditions for inner Lindblad resonance in a spiral galaxy: $2(\Omega_p - \Omega)/\kappa_0 = -1, 0$ and $+1$

Find the values of Ω_p which satisfy each.

4) The Andromeda Galaxy's mass $= 1.2 \times 10^{12}$ solar. Its radius R= 22,000 pc. For a Sun-sized star, one-third of the way to the Andromeda rim, compute: i) the potential V(r), ii) the period of this star, iii) epicycle frequency (κ) if the ratio $\chi = 2$, iv) the specific angular momentum J.

XX. MHD and Solar Physics

Most solar physicists work in what's called the "MHD" or *magneto-hydrodynamic* regime, which numerous plasma physicists have criticized as "taking the easy way out". But is it really? For example, treatments based on the analysis of the indigenous magnetic fields currently occupy a predominant role in analysis of energy stored in sunspot regions. The majority of these are based on MHD (magneto-hydrodynamic) theory wherein low frequency, long scale length and quasi-neutral (electric charge) assumptions are introduced into the plasma physics. Such assumptions lead to a basic set of what we call "MHD" equations, which include:

i) $\partial \rho / \partial t + \nabla \cdot (\rho v) = 0$

ii) $\rho \, \partial V / \partial t + \rho (v \nabla) v = -\nabla p + 1/c \, \mathbf{J} \times \mathbf{B}$

iii) $\mathbf{E} + \mathbf{v} \times \mathbf{B} = 0$

In truth, MHD does represent the end product of successive information loss in the transition from Vlasov theory to two-fluid theory to one-fluid theory and thence MHD. But is the criticism that MHD is "oversimplified" valid in the solar case? Let's first look a bit at the key condition: whether a magnetic field is frozen in or not.

Whether a given solar plasma field is "frozen in" will depend on the magnitude of what is called the "magnetic Reynolds number" (R_m) If $R_m \gg 1$ then diffusion can be ignored, and we have the frozen – field condition and MHD.

The magnetic Reynolds number is commonly expressed:

$R_m = L V_A / \eta$

where L is a typical length scale for the particular solar environment, V_A is the Alfven velocity and η is the magnetic diffusivity or:

$\eta = (5.2 \times 10^7 \ln \Lambda \, T^{3/2})$ m²/s

where the Coulomb logarithm is:

$\ln \Lambda = 9.00 + 3.45 \log (T) - 1.15 \log (n_e)$

For a solar active region of $T = 10^4$ K, one has $\log (T) = 4.00$

For an associated (electron) particle density of $n_e = 10^{10}$ /m³ one has: $\log (n_e) = 10.00$

Then:

$\ln \Lambda = 9.00 + 3.45 (4.00) - 1.15 (10.00)$

$\ln \Lambda = 17.8 - 11.5 = 6.3$

Then the magnetic diffusivity is:

$\eta = 327.6$ m²/s

Now, let L = 10^7 m and V_A = 10^3 m /s, then

R_m = [10^{10} m² /s] / 327.6 m² /s = 3.00 x 10^7

which is a typical value for the conditions referenced, and shows the magnetic field to be frozen into the plasma. Hence, MHD is a valid approach!

Apart from this, the most conspicuous agent of solar plasma containment and topological change is the coronal loop – which is fully consistent with a low beta-plasma regime- since otherwise gas kinetic pressure would predominate and the configuration would be 'open'(Note: β= (½ ρ v²)/ B²/2μ_o where ρ is the fluid density, v the fluid velocity, B the magnetic induction and μ_o the magnetic permeability of free space.)

With this little exercise, we see that the solar physicist is more than amply justified in employing the MHD approach. Indeed, for most applications for assessing sunspot regions, say two fluid theory, it would be far too complicated while achieving little additional insight.

If one is determined to go the 2-fluid route the approach is straightforward: simply take the original MHD equations and treat them as if for two **separate** fluids (ions and electrons) instead of one. Then the first, or charge conservation equation would become, for example:

$\partial \rho_{e,i} / \partial t + \nabla \cdot (\rho_{e,i} \, v_{e,i}) = 0$

where the subscripts (e,i) alert the user to process the equations separately for electrons (e) and ions (i). Is all the extra work worth the information or information quality gleaned? The user would have to determine that for himself in the context of his problem, analysis!

Problems:

1) For a typical solar corona temperature of $T \approx 10^6$ K and a number (fluid) density, ρ of 10^{15} m-3, find the magnetic diffusivity and the magnetic Reynolds number. (Note: the Alfven speed $V_A = 10^3$ ms^{-1})

Take the length scale L= 10^3 m

Solution:

We first need to obtain the Coulomb logarithm:

$\ln \Lambda = 9.00 + 3.45 \log(T) - 1.15 \log(n_e)$

Where $T = 10^6$ K so $\log T = 6.00$.

Similarly, $\log(n_e) = 15$.

Then: $\ln \Lambda = 9.00 + 3.45 \log(6.00) - 1.15 \log(15)$

$\ln \Lambda = 9.00 + 20.70 - 17.25 = 12.45$

Next, the magnetic diffusivity:

$\eta = (5.2 \times 10^7 \ln \Lambda \, T^{3/2}) \, m^2/s$

$\eta = (5.2 \times 10^7 (12.45) (10^6)^{3/2}) \, m^2/s$

$\eta = 12.45 (5.2 \times 10^7 \times 10^9) \, m^2/s$

$\eta = 12.45 (5.2 \times 10^{16}) \, m^2/s = 6.4 \times 10^{17} \, m^2/s$

Last, we find the magnetic Reynolds number:

$R_m = L V_A / \eta$

And we have $L = 10^3$ m, $V_A = 10^3$ ms^{-1} and $\eta = 6.4 \times 10^{17}$ m^2/s, therefore:

$R_m = (10^3 \, m)(10^3 \, ms^{-1}) / (6.4 \times 10^{17} \, m^2/s)$

$R_m \approx 0$ (so the field is not frozen in)

(2) The MHD form of Ohm's law (eqn. (iii)) gives: **E + v X B = 0**

Determine whether the conditions for Problem (1) are in the MHD realm, according to this condition if the full statement of Ohm's law is:

E + v X B = (**J X B**)/ n_e e - $\nabla p_e / n_e$ e + η **J**

(Hint: (**J X B**) only applies in the frozen –in condition and $\nabla p_e / n_e \, e \approx \eta \, \mathbf{J}$ if $\eta >> 0$.)

2. Observational Approaches to Solar Flares:

Most current approaches to flare analysis depend on *the current helicity density*.

The current helicity density can be estimated from the differentials of the transverse field components and how it is related to the force-free parameter, α (Bao and Zhang, *Astronomical Journal*, Vol. L43, p. 496, 1998):

$$H_z(c) = [\partial B(x)/\partial y - \partial B(y)/\partial x] B_z$$

An illustration of this is provided below in solar magnetic maps:

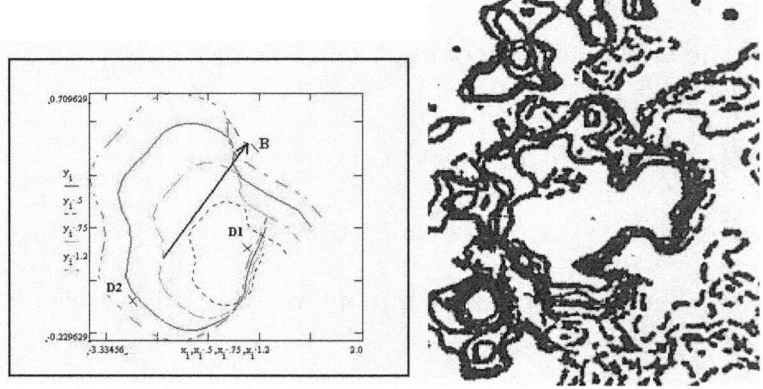

Fig. 1 Left: Model magnetogram to estimate $\partial B(x)/\partial y$, and $\partial B(y)/\partial x$. Right side is a contour sketch of the region from the actual Mt. Wilson magnetogram for Nov. 5, 1980. The "bird's beak" defines the approximate magnetic inversion (neutral) line.

This gives an excellent idea of how solar observations can be interpreted to give useful results. In the diagram at left, the visual range presents a model photospheric vector magnetogram obtained between 12h 00 GMT and 13h 30 m GMT on a particular date. Each increment of 0.25 along

DELTAY ("Delta Y") or DELTAX ("Delta X") denotes a change of 250 km while each contour difference is a separation of 250 G. The B-vector direction has been estimated as shown, and the components B(y) and B(x) can be worked out, along with the changes, e.g. $\partial B(x)/\partial y$ and $\partial B(y)/\partial x$, say from one iso-contour to the next.

From the diagram above, let us work out $\partial B(x)/\partial y$ and $\partial B(y)/\partial x$. We first note that $\partial B(x)$ effectively crosses three contours, or (3 x 250 G) = 750 G. The separation dy (or ∂y) from the vertical axis amounts to ≈ 1.0 unit(s) or 4 x 250 km ≈ 1000 km. Thus:

$\partial B(x)/\partial y$ = (750 G)/ 1000 km = 0.75 G/ km

which is well over the limit that Severny et al (A. .B. Severny, N.N. Stepanyan, and N.V. Steshenko.: 1979, in R.F. Donnelly (ed.), *Solar-Terrestrial Predictions Proceedings*, Vol. 1) defined for probable flares (e.g. 0.1 G/ km). In a similar way, we find for $\partial B(y)/\partial x$:

$\partial B(y)/\partial x$ ≈ (750 G) / (3.5 x 250 km) ≈

750 G/ 875 km ≈ 0.85 G/ km

We are now in a position to estimate the current helicity density.

$H_z(c) = [\partial B(x)/\partial y - \partial B(y)/\partial x] B_z$

≈ [0.75 G/km - 0.85 G/km] 350 G

$H_z(c)$ ≈ - 35 G^2/ km

and the negative sign indicates that *the force-free parameter* α is also negative, and will have magnitude:

$\alpha \approx -[\partial B(x)/\partial y - \partial B(y)/\partial x] / B_z$

$\approx -[0.75 \text{ G/km} - 0.85 \text{ G/km}] / 350 \text{ G}$

$\alpha \approx -(10^{-4} \text{ m}^{-1}) \text{ G} / 350 \text{ G} \approx -2.8 \times 10^{-7} \text{ m}^{-1}$

The sign fixes the solar hemisphere in which the region occurs. Thus the sign of helicity will be positive or negative, depending on what is known as the *"hemispheric helicity rule"*. That is, the force-free α characterizing each active region will have a tendency to be (+) in the southern solar hemisphere, and (-) in the northern solar hemisphere.

Whether extra energy is stored in a localized solar active region depends on whether we call the "force-free" condition applies or not. This condition is:

$\nabla \times (B) = \alpha (B)$

The force-free parameter is usually expressed (where $\mu_0 = 4\pi \times 10^{-7}$ H/m) :

$\alpha = \mu_0 J_z / B_z$

where J_z is the vertical current density (measured in amperes per square meter, e.g. A m^{-2}) and B_z is the normal component of the magnetic induction at the photosphere.

For the special case, $\alpha = 0$, then $\mu_o J_z = 0$ and we obtain a "current free" configuration for which there is no residual energy to be extracted from the field, e.g. for flares. This is also called a "potential" field. For any force-free field for which $\alpha > 0$, magnetic free energy is available for flares.

Problem:

Refer to the illustration of the previous problem, to get the current helicity density and the force-free parameter α. Is that field current free or does it have excess magnetic energy? If the latter, then estimate the vertical current density J_z if $B_z = 0.1T$.

Solution:

Since $\alpha \approx -2.8 \times 10^{-7}$ m^{-1} then clearly, $\alpha \neq 0$ so the field cannot be current-free or potential. Hence, excess magnetic energy must be stored in the field.

The vertical current density can then be estimated from:

$\alpha = \mu_o J_z / B_z$

Re-arranging: $J_z = \alpha B_z / \mu_o$

$J_z = (2.8 \times 10^{-7}$ m$^{-1})(0.1T)/(4\pi \times 10^{-7}$ H/m$)$

$J_z = 0.02$ A m^{-2}

Where values on the order of 0.001 A m^{-2} are definitely in the flare regime.

The power, energy of solar flares is also estimated from the soft x-ray records such as shown below from a sample over November, 1980.

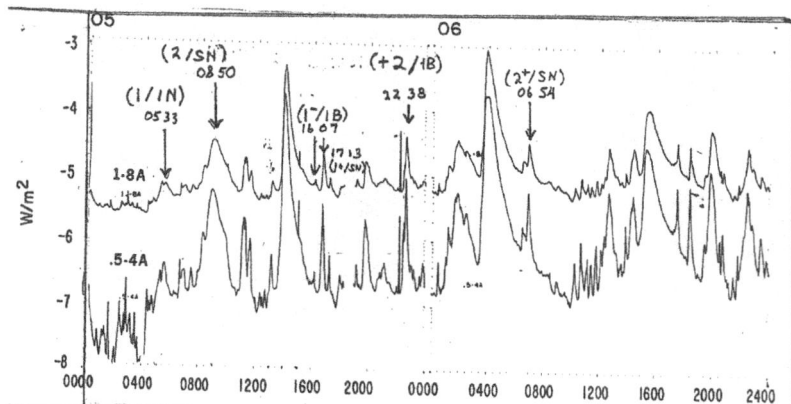

Fig. 2. GOES- Soft x-ray flux record over Nov. 5-6, 1980 with assorted flares identified.

The estimation of flare energy from the soft x-ray record is fairly straightforward and entails multiplying the SXR flux (left axis) of the "half-power" points by the time duration (horizontal axis) then by the recorded flare area in square meters.

As an example, consider the flare occurring at 04h 30m UT on Nov. 6 with an estimated half-power point flux of $F = 10^{-5}$ W m^{-2} and a duration $t \approx 3h \approx 10800s$. If the records show it has a flare area of 10^{23} m^2 then:

Flare energy $\approx (10^{-5}$ W m$^{-2})(10800s)(10^{23}$ m$^2)$

Flare Energy $\approx 1.1 \times 10^{23}$ J

Recall for this computation that Watts (W) = J/s or joules per second, hence the units leave J at the end so we know its energy.

The power of the flare can also be obtained, simply by dividing the energy by the time of duration, so:

Power = Energy / time = $(1.1 \times 10^{23}$ J$)/$ 10800s

Flare power = 10^{19} W

Other quantitative assessments are possible but they require that specific models be applied (say for 'double layers') or that other ancillary measurements be known (say the magnetic flux, φ = BA, where A is the area of the spot, say, and B the magnetic induction).

If the electrical resistance R, associated with a pre-flare system is known, then the current I can be estimated and the region assessed for a flare. Thus, if R is known and we know (from basic physics) that:

$P = I_o^2 R$

Then: $I_o = [P/R]^{1/2}$

where I_o is the pre-flare current. If I_o is then known it is possible to obtain the voltage drop (V(t)) in a double layer (assuming that model applies) since:

$P = I V(t)$

(Note: Generally double layers are regarded as unphysical because of their small dimensions)

If both V(t) and I are then known, it is feasible to obtain the change in the system's inductance (dL/dt) since:

$V(t) = I \, dL/dt$

So that: $dL/dt = V(t)/I$

In other words, if only one more parameter is quantitatively known or can be inferred/estimated, it is possible to extract much more information on the whole system and discern whether or not a flare is likely!

Problem:

1) From the soft x-ray record in Fig. 2, estimate the energy of the flare occurring at 08h 50m on Nov. 5. Use this to also obtain the power of the flare of Solar Geophysical data for the date and time estimates the flare areas at 6.1×10^{22} m^2. If the typical resistance found in the region before the flare is estimated to be 0.0047Ω find the current I associated with it.

Solution:

The soft x-ray record shows a flare whose mid-power points yield a flux of 10^{-5} W m^{-2} and time duration of ≈ 9000s. Then the flare energy is:

$E \approx (10^{-5} \text{ W m}^{-2})(10800\text{s})(6 \times 10^{22} \text{ m}^2)$

Flare energy $\approx 6.4 \times 10^{21}$ J

Power $= (6.4 \times 10^{21} \text{ J})/10800\text{s} = 6.0 \times 10^{17}$ W

If R = 0.0047Ω then the current is found from the power –current equation:

P = I_o^2 R

So: I_o = $[P/R]^{1/2}$ = $[6.0 \times 10^{17}$ W/ 0.0047Ω $]^{1/2}$

I_o ≈ 1.1 x 10^{10} A

Other Problems:

1) Examine the magnetogram below for a sunspot region.

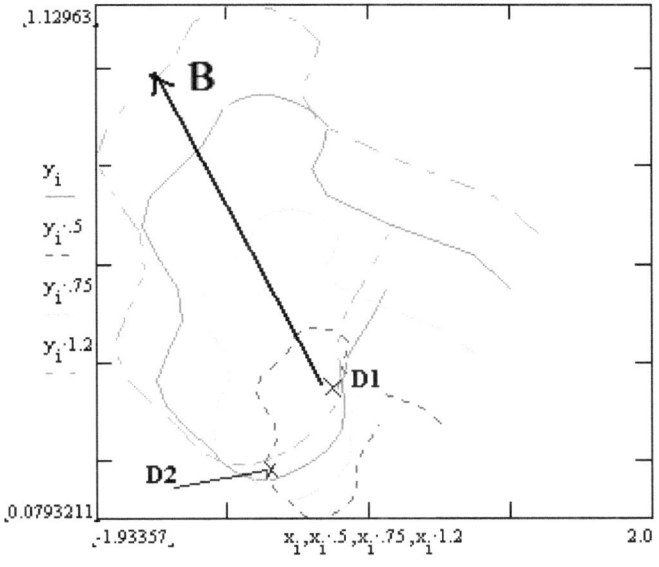

Note that each increment along "Delta Y" or "Delta X" denotes a change of 250 km while each contour difference is a separation of 250 G. The B-vector direction has been indicated as shown.

From this information and the contour map work out the gradients $\partial B(x)/\partial y$ and $\partial B(y)/\partial x$.

Then: Obtain the current helicity density, and the force-free scale factor applicable to the region.

2) Given the force-free scale factor obtained in (1) and a vertical current density $J_z \approx 0.012$ A m^{-2} estimate the magnitude of the magnetic component B_z.

3) Melrose, 2004 (*Conservation of Both Current and Helicity in a Quadripolar Model for Solar Flares*, in Solar Phys.,) has proposed a way to measure the current from *the average* of the force-free parameters (α_I) obtained over the loop cross section as:

$I_o = [\alpha_I \varphi_i]/\mu_o$

where φ_i is flux of the ith loop is computed from the equipartition value of the magnetic field observed and the observed loop cross sectional area, A.

Assume the force free scale factor obtained in (1) is *also the average, of the* α_I and the equipartition value of the magnetic field B is 0.0084T. If the loop spanning flux centers D1 and D2 in the contour is found to have area A = 1.4 x 10^{13} m^2 then estimate *the current* in the system. If it was a double layer system and the observed flare power was 8.0 x 10^{17} W, what would be the expected voltage drop over the double layer? Estimate the resistance R of the double layer system.

XXI: Introduction to Special Relativity

In the popular mind, at least, the word "relativity" usually conjures up visions of space travelers returning youthful to Earth after journeys of many decades (measured on Earth). Of course, there is much more to relativity than this. As a matter of fact many aspects of relativity as in the case of motions, aren't new at all. Basically, it merely entails the assertion that the laws of physics appear to be the same in given reference frames or coordinate systems. This is made more accurate in Einstein's special relativity by referring to *inertial reference frames.*

Though Henri Poincare came close to discovering the principles of relativity, it was Albert Einstein first and foremost who ruthlessly and relentlessly pursued the basic principles to their utterly logical conclusions, including that lengths shrink as velocities tend toward the speed of light, c, and time slows or "dilates".

Like Ernst Mach before him, Einstein adopted the view that in considering two objects in relative motion, it is futile and meaningless to attempt to decide which object is "really" in motion and which is at rest. If you are in a Jumbo jet traveling at a speed of 900 km/hr relative to the Earth, it makes no difference whether someone says you are moving at that speed, or the Earth is moving at that speed. In either case, the operation of the laws of physics in your jet and on the Earth will be the same. Balls will still drop, and you will still fall off your chair if not careful. There exist no absolute frame of reference to contradict you.

The Special Theory of Relativity, interestingly, may be said to have had its origins in the null result of an experiment the primary aim of which was to detect relative motion. This experiment was first carried out by the American physicist Albert A. Michelson in 1881, and subsequently repeated in 1887, with the help of Edward W. Morley. The experiment has thus come to be known as "the Michelson-Morley experiment".

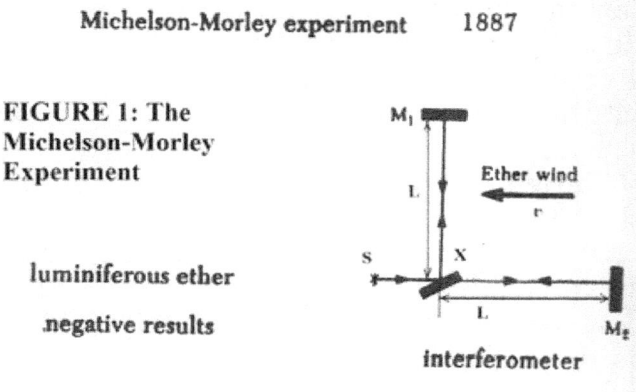

FIGURE 1: The Michelson-Morley Experiment

luminiferous ether

negative results

The basic idea was to time the transits of light in two distinct directions: perpendicular to the Earth's orbital motion, and parallel to the orbital motion. This was to be accomplished by using an arrangement of mirrors and light beams such as depicted in Fig. 1. A difference in light velocities (transit distance/ time) would reveal itself by a delicate interference pattern formed by two separate beams after rejoining each other.

The implication of a difference in velocities would mean the confirmation of a remarkable entity called "the Ether". In effect, if light represents waves propagating through the Ether, the velocity of light as recorded by instruments on Earth's surface must be distorted by the motion of the Earth through space. An analogous principle is that a swift river must retard a swimmer's combined upstream and downstream speeds more than his cross-current speed. Similarly, a large difference in light velocities (along the two different paths) should show Earth is moving rapidly through the Ether, while a small difference would show it's moving slowly.

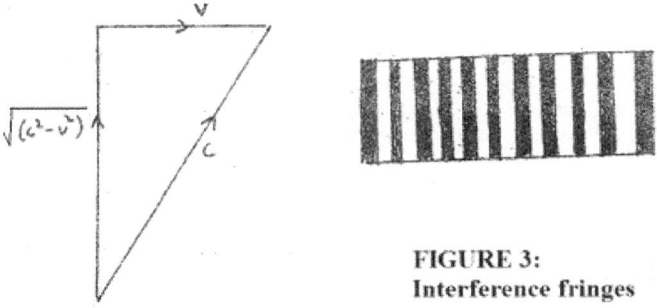

FIGURE 3:
Interference fringes

FIGURE 2

Astonishingly, on each occasion the experiment was conducted the result was virtually negative. There was no evidence of any Ether flowing past the Earth in any direction. Any minor deviations fell within the purview of the experimental errors.

Imagine the apparatus in Fig. 1 to be moving with velocity v toward the right, then only in the event of *a null result* should there should there also be a relative velocity of the Ether of magnitude v to the left. To get a positive result the apparatus needs to move a speed v relative to the Ether. The round trip time for a light beam following path X-M2-X is:

$t_1 = 2Lc/(c^2 - v^2)$

Meanwhile, the light beam traveling from X to M1 must have a component of velocity v along X-M2, relative to the hypothesized Ether or it will not strike the mirror at M1. Since the velocity of the light relative to Ether is c, subtracting the preceding component leaves a velocity of $(c^2 - v^2)$ e.g. see Fig. 2. The same is true for the return journey to the total time t2 for the path X-M1-X is:

$t_2 = 2L/(c^2 - v^2)^{1/2}$

Now, the recombination of two light beams (originally split in two) will produce interference fringes as depicted in Fig. 3. Any difference in the times taken to traverse the paths will show up as a shift in the position of the bright and dark fringes since it will indicate different path lengths.

Hence, a difference in time delta t is equivalent to a path difference:

$c(\Delta t)/\lambda (d)$

where d = the width of one fringe and lambda is the wavelength of the light (taken to be 6×10^{-7} m).

From this, the time difference between the two beams can be computed from:

$\Delta t = t_1 - t_2$

$= 2L/c \{ 1/(1 - v^2/c^2) - 1/(1 - v^2/c^2)^{1/2} \}$

This equation can be approximately expressed as:

$\Delta t \approx L/c \, (v^2/c^2)$

Since $v \ll c$, this corresponds to a fringe shift of: $\Delta d - c \, (\Delta t)/ \lambda = L \, c^2 / \lambda \, (c^2)$ The measurement of the fringe shift is accomplished by rotating the whole apparatus through 90 degrees. The effect of this is to interchange the arms, X-M1 and X-M2, thereby reversing the sign of the fringe shift so that an overall shift of 2 (delta d) should be observed. For the Michelson-Morley experiment of 1887, L = 11 m, v = 0.0001c. Using the path difference equation for Δd, we arrive at: Δd = 0.183 or, approximately 0.2 fringe. The overall shift expected was: 2 x (Δd) = 2 x (0.2) = 0.4 fringe But the largest shift *actually detected* was only 0.01 fringe within the experimental error.

This null result flabbergasted physicists of the time. They were simply unable to conceive that light required no medium within which to propagate. Hence, it's not difficult to see why so many clung to the Ether McGuffin for so long, even after it was disproven. Thus, fanciful and elaborate schemes were thought up to explain the null result, much like today's intelligent design proponents have confected fanciful explanations to try and disavow Darwinian evolution.

For his own part, Michelson naturally assumed that the local ether had to be adhering to the Earth, travelling with it through space. All other scientists were incredulous that an orbital velocity of 30 km/s in relation to the Sun could not generate the tiniest ether "breeze". To many, the situation was not unlike a ship maintaining constant speed and direction in the sea, irrespective of current changes.

Problems

1. In the Michelson-Morley experiment, the length L of each arm of the interferometer was 11 meters. Sodium light of wavelength 5.9×10^{-7} m (590 nm) was used. The experiment would have revealed any fringe shift > 0.005 fringe.

What upper limit does this place on the Earth's velocity through the supposed Ether?

2. Using Fig.1, say the time of travel to the right is: $t(r) = L/(c - v)$ and the time of travel to the **left** is $t(L) = L/(c + v)$.

a) Find the "total time of travel" by adding both left and right contributions.

b) Find the time consumed for "a half-trip".

c) Find the time consumed for a round trip.

d) Add the two "half trips" and what do you obtain?

e) Why does this not agree with the value obtained for (a)?

f) Work out what these times consumed would be for a trip to Proxima Centauri., using an ion –powered craft able to travel at $v = 0.1c$.

Solutions:

Given $L = 11$ m and $\lambda = 5.9 \times 10^{-7}$ m

limit for lowest resolution: $2 \Delta d = 0.005$ fringe

Then:

$d = (0.005 \text{ fringe})/2 = 0.0025$ fringe

and since:

$\Delta d = Lv^2/(c^2)$

$v = \{(\Delta d)(\lambda) c^2 / L\}^{1/2}$

$v = \{(0.0025)(5.9 \times 10^{-7} \text{ m})(3 \times 10^8 \text{ ms}^{-1})/11 \text{ m}\}^{1/2}$

$= 3.5$ km/s

2. We are going along (parallel) or opposed (anti-parallel) to, the "ether wind" direction from the diagram and this is *horizontal* so designate it by direction x, e.g.

$t(\text{total } x) = L/(c + v) + L/(c - v)$

The common denominator is:

$(c - v)(c + v) = c^2 - cv + cv - v^2 = c^2 - v^2$

Then $t(\text{total } x) = 2Lc/(c^2 - v^2)$

$t(\text{total } x) = 2L/c \, (1 - v^2/c^2)^{-1}$

b) The time consumed for "a half-trip".

half trip time = (t(total x))/2 = ½ [$2Lc/(c^2 - v^2)$]

(t(total x))/2 = $Lc/(c^2 - v^2)$

But recall that the interferometer has parallel and perpendicular components so that total time registering both is:

T(total x+y) = $t_1 + t_2$ where:

$t_1 = 2Lc/(c^2 - v^2)$

$t_2 = 2L/(c^2 - v^2)^{1/2}$

So:

$t_1 + t_2 = 2Lc/(c^2 - v^2) + 2L/(c^2 - v^2)^{1/2}$

$= 2L\{c + (c^2 - v^2)^{1/2}\}/(c^2 - v^2)$

so $T/2 = L\{c + (c^2 - v^2)^{1/2}\}/(c^2 - v^2)$

c) Find the time consumed for a round trip.

Round trip is $2(t_1 + t_2)$

$= 2(2L\{c + (c^2 - v^2)^{1/2}\}/(c^2 - v^2))$

$= 4L\{c + (c^2 - v^2)^{1/2}\}/(c^2 - v^2)$

d) Add the two "*half trips*" and what do you obtain?

We get:

$2(T/2) = 2L\{c + (c^2 - v^2)^{1/2}\}/(c^2 - v^2)\}$

e) Why does this not agree with the value obtained for (a)?

It doesn't agree because each summing (or halving) as computed above takes into account *differing directional components* of velocity with respect to the interferometer. These need not be equal!

The Michelson –Morley Experiment:

In a last ditch effort to satisfactorily explain the riddle of the null result (from the Michelson-Morley experiment), a fantastic idea was put forward by George F. Fitzgerald in 1890. Using the analogy of a rubber ball which is deformed upon striking a wall, Fitzgerald conceived that the ether would distort matter.

This distortion would take the form of a contraction of length in the direction of the motion through the ether. Such a contraction would explain the null result of the Michelson -Morley experiment. That is, the arm L, of the apparatus, moving against the ether would be shortened by "ether pressure" just enough to compensate for the slowing down of light by the ether wind.

A similar but more mathematical theory was

worked out by the physicist Hendrik A. Lorentz, who expressed the length contraction in the direction of motion by:

$$L' = L[1 - v^2/c^2]^{1/2}$$

This famous equation became known as the "*Lorentz-Fitzgerald contraction*".

Understandably, the hypothesis of Lorentz and Fitzgerald gradually gained general acceptance, except for a Swiss Patent clerk, who refused to be deluded by such a contrived idea (based as it was on absolute motion). The clerk's name was Albert Einstein, born in Ulm, Bavaria, in 1879.

Einstein had graduated as a physics major, but he managed to produce such a poor impression as a teacher that he was dismissed from three teaching jobs in succession. Having been reduced to a 'hand-to-mouth' existence, he was lucky to find a job processing patent applications in Bern, Switzerland. Fortunately, he also has a leisurely schedule that permitted him to while away a good many hours contemplating space, time and energy. (A good thing his time was nearly a century ago, otherwise - today- he'd be accused of "gold bricking"!)

After much thought, Einstein was forced to conclude that motion is never observable as motion with respect to space and that there is no basis for the introduction of "absolute motion". In Einstein's mind, the only kind of motion was relative rest perceived from different viewpoints (e.g. "reference frames"). Einstein called this insight the Principle of Relativity.

According to Einstein's Principle of Relativity: *"All the laws of physics are the same in all inertial reference frames."*

It's instructive at this point to commence a quantitative approach to see exactly how Einstein reasoned. This necessitates we first get accustomed to the idea of a coordinate system. Basically, the purpose of any coordinate system is to enable us to identify a particular point in space. The standard procedure (for 3-dimensional space) is to take three mutually perpendicular axes with coordinates in x, y and z, representing distanced to where the axes meet at the origin, or 0. In special relativity such coordinate systems are typically designated: S, S', S" etc. or alternatively, S1, S2, S3.

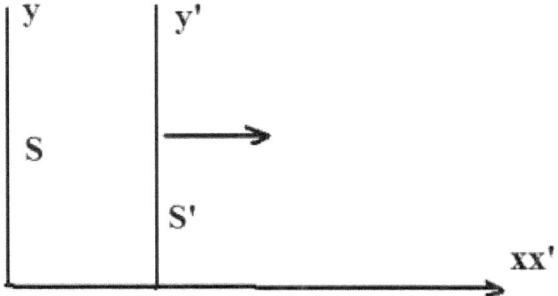

FIGURE 1: Coordinate systems S and S' with relative motion in x-direction.

In Fig. 1 we have two coordinate systems, S and S', with S having coordinate x, y and S' having coordinates x', y'. (We confine the systems to 2 dimensions here for initial simplicity.) Thus, S' has its origin at 0' and S at 0. In Fig. 1, S' is shown moving with constant velocity v in the x - x' direction (which obviously coincide for S and S').

Keep in mind though we arbitrarily assigned S' as moving this depends on which coordinate system is taken to be at rest. An observer attached to system S may very well consider himself moving and S' stationary. This would resemble the well-known example of a passenger in a stationary train observing a train parallel to his through his window and deducing he is moving, though it is actually the train parallel to his. The key point is that the systems S and S' have a constant relative velocity. Such coordinate systems occupy a special place in relativity and are called "inertial reference frames" or "inertial coordinate systems". Their primary feature is that they lack and acceleration of one to the other.

Now, think about this carefully: if the origins O and O' coincide at time t = 0, and we are observers in S', then we will see o' move along OX with velocity v. Then the two sets of coordinates, representing the same point (in S and S') are related by:

$x' = x - vt$ and $y' = y$

If we chose we could append a 3rd axes (z) i.e. coming out of the page, and have also:

$z' = z$

The preceding equations describe the "Galilean transformation". From these it's easy to obtain velocities by differentiating with respect to t:

$dx'/dt = dx/dt = v$

$dy'/dt = dy/dt$

$dz'/dt = dz/dt$

(Note here that never once did we write t' = t since the notion of an absolute, universal time was the very cornerstone of Newtonian theory!)

We now start with the preceding transformations and see how Einstein reasoned. Reference may be made to Fig. 2 which displays a three-dimensional perspective of the relative motion and hence is a bit more complicated. Our aim in using it is to find an alternative to the Galilean transformations - which obviously can't be correct for all situations.

For example, according to the Galilean transformations, a pulse o flight sent out from S' would move with the velocity c, the speed of light, as measured from S', but with the velocity (c + v) measured from S. But this directly contradicts the demonstrated fact (Michelson-Morley) that the speed of light is always constant when computed from one reference frame relative to another.

Einstein thus began afresh by using not only the Principle of Relativity (just stated, i.e. the laws of physics are the same in all *inertial* reference frames)) but also:

The speed of light is always found to have the same value no matter what the motion for the source or the observer.

From these two postulates, Einstein deduced a

number of surprising results which would have been totally unacceptable to a more conservative mind.

Start then with the two systems depicted in Fig. 2 which are coincident in space at the instant a flash bulb (say) is set off when the origins coincide. The observer S sees S' moving in the x-direction with the velocity v and observer S' sees S moving with the same velocity in the opposite (-x) direction.

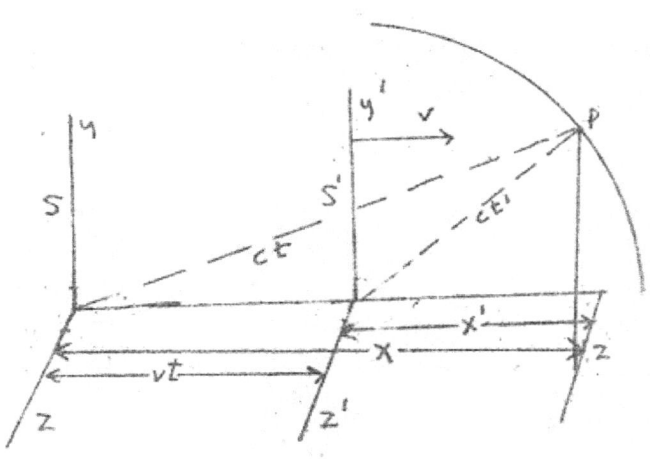

FIGURE 2: RELATIVE MOTION OF S' AND S referred to space-time event

Both observers see the flash of light travel away from the origin with the same velocity c. The distance the light travels in the S system during time t, is ct. This must be true in any direction. If the light spreads out equally in all directions, then by the end of time t it has extended to fill a sphere whose radius is r, and we can write:

(1) $r^2 = x^2 + y^2 + z^2 = c^2 t^2$

The exact same equation holds in the S' system since the flash was set off when O coincided with O. Thus, while the 2 systems are separating from each other, the observer in S' also sees the light fill a spherical shell in his own system, so writes (for r'):

(2) $r'^2 = x'^2 + y'^2 + z'^2 = c^2 t'^2$

Note the velocity of light is the only thing that's the same in both systems. The two preceding equations thus describe a point lying on the spherical shell. Even though they describe the same point in space, the observer in S sees the point at position (x, y, z, t) and the observer S' at position (x', y', z', t').

Subtracting equation (2) from equation (1) and transposing terms:

$x^2 + y^2 + z^2 - c^2 t^2 = x'^2 + y'^2 + z'^2 - c^2 t'^2$

We now look for a transformation similar to the Galilean transformation, but which will allow c to be the same in both S and S'. Since the y and z coordinates of the position are not affected by the motion in the x-direction we can say y' = y and z' = z. For the x-coordinate, we try a transformation of the form: x = a(x' + vt') and x = a(x - vt), where a is an invariant determined by the two fundamental postulates (i.e. the same quantity is used in going from x to x' as from x' to x).

Further, we expect a to depend on the velocity v in such a way that it becomes equal to 1 when v becomes very small compared with the speed of light. When this happens, the x and x' transformations become the same as the ordinary Galilean transformations.

So we begin by using $x = a(x' + vt')$ to solve for t' and obtain:

(3) $t' = 1/v \, (x/a - x')$

For x' above, we now insert the value for x' (e.g. $x' = a(x - vt)$):

(4) $t' = 1/v(x/a - ax - avt) = at - x^2(a^2 - 1)/va$

Similarly, we find for t:

(5) $t = -at' + x'(a^2 - 1)/va$

If we now substitute $x' = a(x - vt)$ and equation (4) into the right hand side of equation (2), we obtain:

(6) $x^2 + y^2 + z^2 - c^2 t^2 = a^2(x - vt)^2 + y^2 + z^2$

$= c^2[at - x/v \, (a^2 - 1)^2/a]$

Now, re-arrange terms and cancel the z and y terms, which are the same on both sides of the equation, to get:

(7) $x^2 - c^2 t^2 = [a^2 - (a^2 - 1)c^2/ a^2v^2]x^2$

$+ 2[(a^2 - 1)c^2/v^2 - a^2] xvt - (c^2 - v^2)a^2t^2$

If the preceding is to hold true for any value of x and t, each term on the left side must equal each term on the right. Since there are no terms with the combination xt on the left, the xt term on the right must be zero. This means:

(8) $(a^2 - 1)c^2/ v^2 - a^2 = 0$

Solving for a:

(9) $a^2 = 1/ (1 - v^2/c^2)$ and $a = [1/ (1 - v^2/c^2)]^{1/2}$

Finally:

(10) $(a^2 - 1)/a = v^2/c^2/ [1/ (1 - v^2/c^2)]^{1/2}$

Substituting the preceding into our x, x' transformation equations and equation (3), we arrive at the following transformations to replace the Galilean:

(11)

$x = x' + vt'/(1 - v^2/c^2)]^{1/2}$

and

$x' = x - vt/ (1 - v^2/c^2)]^{1/2}$

(12) y = y' and y' = y

(13) z = z' and z' = z

(14) $t = t' + x'v/c^2/[(1 - v^2/c^2)]^{1/2}$

and

$t' = t - xv/c^2/[(1 - v^2/c^2)]^{1/2}$

Equations (11)- (14) are known as the **Lorentz transformation** or the Lorentz-Einstein transformation.

Note the important feature is that the *time* must be given as well as the position, because the respective clocks in S and S' will cease to read identical times after they have parted from one another. This is the significance of (14) in the above set. Also, the fact that time is given the same importance as space (i.e. as another dimension) shows there's nothing special or mystical about "the fourth dimension".

Problems:

1- Given that $x' = 1/a \, (x - vt)$ and $t' = 1/a \, (t - vx/c^2)$, derive similar equations for x and t in terms of x' and t'. (Recall: $1/a = (1 - v^2/c^2)^{1/2}$)

2- An event in space-time occurs at x' = 60 m, t = 8 x 10^{-8} s, in a frame S' (y' = 0, z' = 0). The frame S' has a velocity of 0.6c along the x-direction with respect to a frame S. The origins O and O' coincide at time t = t' =

0. Find the space-time coordinates of the event in S.

3- Suppose an astronaut is traveling at 0.9c in a space ship with respect to the Earth. How long a time interval will his clock indicate when the Earth has revolved once around the Sun? (Take the duration of one standard revolution of Earth around the Sun to be 365 ¼ days.)

Solutions :

1) Given $x' = 1/a (x - vt)$ and $t' = 1/a (t - vx/c^2)$,

Then: $x' = x/a - vt/a$ and $t' = t/a - vx/ac^2$

and: $x' + vt/a = x/a$ and $t' + vx/ac^2 = t/a$

so:

$x = a(x' + vt/a)$ and $t' = a(t' + vx/ac^2)$

finally: $x = a(x' + vt)$ and $t = a(t' + vx/c^2)$

2) We have: $x' = 60m$, $t' = 8 \times 10^{-8}$ s and $y' = y$, $z' = z$

$v = 0.6c = 1.8 \times 10^8$ m/s

Then:

$x = [60m + (1.8 \times 10^8 \text{ m/s})(8 \times 10^{-8} \text{ s})]/ (0.64)^{1/2}$

$x = [60m + 14.4m]/ 0.8 = 74.4m/0.8 = 93m$

and t =

[(8 × 10⁻⁸ s) + (1.8 × 10⁸ ms⁻¹)(60m)/(3 × 10⁸ ms⁻¹)²/0.8

t = 2.5 × 10⁻⁷ s/ 0.8 = 2.33 × 10⁻⁷ s

The space time coordinates are: (93 m, 2.33 × 10⁻⁷ s)

3) The problem requires no relative motion defined specifically in the x-direction so the equations:

$t = t' + x'v/c^2/(1 - v^2/c^2)^{1/2}$

and

$t' = t - xv/c^2/(1 - v^2/c^2)^{1/2}$

are immediately simplified by the terms in x becoming zero, so:

$t = t'/(1 - v^2/c^2)^{1/2}$

and

$t' = t /(1 - v^2/c^2)^{1/2}$

Here: t = time passage on Earth clock

and t' = time passage on astronaut's clock

For t = 1 Earth year = 365 ¼ days:

$t' = (365 ¼ \text{ days})/ [1 - (0.9c)^2/c^2]^{½}$

$= (365 ¼ \text{ days})/(1 - 0.81)^{½}$

$t' = (365 ¼ \text{ days})/0.436 = 837.7 \text{ days}$

This is the time elapsed on the astronaut's clock when the Earth has made one revolution equal to 365 ¼ days. In other words, each of his days is roughly equal to 2.29 Earth days. Hence, his clock is obviously running *slower* than the Earth clock.

A more intuitive way to look at the result would be in terms of the time transformation:
$t = t'/ (1 - v^2/c^2)^{½}$ - or asking how much time elapses on an Earth clock for each year elapsed on the astronaut's? The result will be found to be 2.29 years, or in other words his Earth counterparts are aging 2.29x faster.

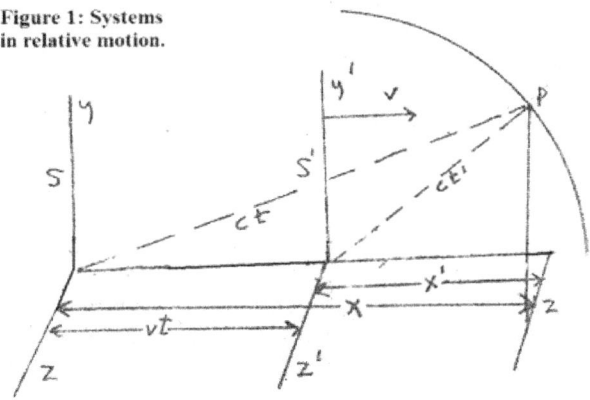

Figure 1: Systems in relative motion.

Let's now consider the implications of the Lorentz-Einstein transformation and refer again to the original diagram (Fig. 1). Recall:

(11) $x = x' + vt'/(1 - v^2/c^2)^{1/2}$

and

$x' = x - vt/(1 - v^2/c^2)^{1/2}$

(12) $y = y'$ and $y' = y$

(13) $z = z'$ and $z' = z$

(14) $t = t' + x'v/c^2/(1 - v^2/c^2)^{1/2}$

and

$t' = t - xv/c^2/(1 - v^2/c^2)^{1/2}$

Suppose there's a clock located in system S' and an observer in system S sees this clock moving with velocity v. At any time t, the position of the clock with reference to the S −observer's system is given by $x = vt$. If the length of time between two ticks is of this clock is T' in the S' system, the transformation to the S system (assuming $x' = 0$) makes the time interval T appear to be:

$T = T'/(1 - v^2/c^2)^{1/2}$

Since T > T' then it seems to Observer S his clock is running slower than it does to S'. This applies not only to the clock but to all physical processes that depend on time, e.g. the vibrations of electrons in atoms, rates of chemical reactions, biological processes (heart beat, respiration) etc. In effect it appears to the observer in S that his counterpart in S' is living at a slower rate than he is. However, to the observer in S' it is the observer in S who seems to be living at a slower rate.

This is a paradoxical result but one which can't be escaped if we carry special relativity to its logical conclusion as Einstein did. So long as the two systems have a constant relative motion, we cannot say that one is moving and the other standing still, or that one's clock is moving slowly and the other quickly.

A number of subtle consequences arise out of this. For one thing, the notion of *simultaneous events*, i.e. for observers in two different reference frames, is no longer tenable. Einstein, in fact, seems to have been the first human being to recognize that simultaneity between two events is a provincial illusion. In his landmark (1905) paper on special relativity ("Does the Inertia of a Body Depend on Its Energy Content?') he remarked:

"We have to take into account that all our judgments in which time plays a part are always judgments of simultaneous events. If, for instance, I say 'That train arrives here at 7 o'clock' I mean something like this: 'The pointing of the small hand of my watch to seven o'clock and the arrival of the train are simultaneous events."

Einstein agreed that observing such simultaneity was a reasonably accurate way for a person holding a watch to tell the time of an event happening **next to the watch**, but insisted that on principle the method couldn't be relied upon for timing an event far away from the watch, especially by someone moving in relation to the other things involved.

As an illustrative example, consider Fig.2 which features an enormously long train (defined as system S') with an observer 'Bob' at its very center. (E.g. if the total length of the train is L, he is at L/2).

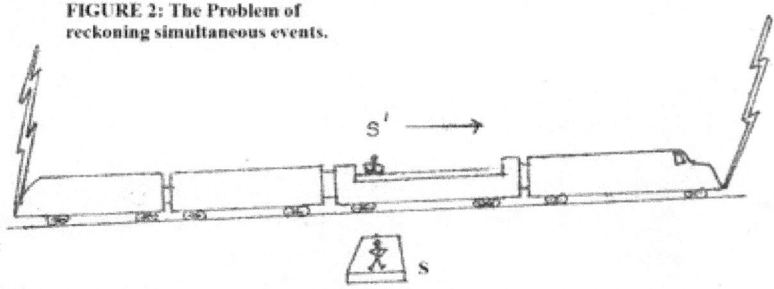

FIGURE 2: The Problem of reckoning simultaneous events.

Two bolts of lightning then strike the extremities of the train, and do so simultaneously according to Bob's watch. Bill, however, is situated on a stationary platform as shown – defined in system S- at the exact midpoint between the strikes and swears that he recorded the rear flash a sizeable fraction of a second earlier than the forward flash.

Who is correct? A non-relativistic thinker would undoubtedly incline towards the view of Bill, the stationary observer. The argument here is that Bob's simultaneity is only apparent, not actual since the train would have carried him closer to the forward flash by the time he received their simultaneous light from points that were, by then, no longer equidistant from him.

For a relativist, however, it is just as true to say the

train was standing still and the ground sped by it, so the ground observer (Bill) could have been carried beyond the equidistant point as easily as Bob.

In Einstein's cosmic viewpoint, the lightning flashes (and their midpoint) are free to be considered an integral part of the train OR Earth, OR any other frame of reference. Each and every observer may select his own viewpoint. Any viewpoint is true and right, none is wrong. This example embodies the essence of relativity, that is, that time is truly relative and not absolute as Newton had believed.

As an illustration of a type of experiment that attempts to reckon simultaneity, consider Fig. 3.

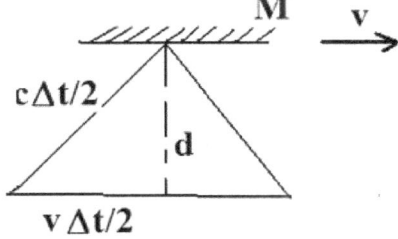

Fig. 3: Simultaneity experiment

Here a light pulse is directed at a moving mirror M traveling with velocity v to the right. By the time the pulse reaches the mirror it has moved a distance $D = v(\Delta t)/2$ horizontally.

According to an outside observer, if light is to hit the mirror it must depart at an angle to the vertical as shown.

By Pythagoras' theorem (square of the hypoteneuse equals the sum of the squares of the other 2 sides):

$(c (\Delta t)/2)^2 = (v (\Delta t)/2)^2 + d^2$

This implies:

$\Delta t = 2d/(c^2 - v^2)^{1/2} = 2d/c[(1 - v^2/c^2)^{1/2}]$

But, for the stationary observer:

$\Delta t' = 2d/c$

which implies:

$\Delta t = \Delta t'/(1 - v^2/c^2)^{1/2}$

We call delta t' the "proper time" or that time interval between two events as measured by an observer who sees the events at the same place. It's always the time measured by a single clock at rest in the frame.

Example Problem:

The period T of a pendulum is measured to be T= 3.0 s in the inertial frame of reference of the pendulum. What is the period of the pendulum when measured by an observer moving at a sped of 0.95c with respect to the pendulum?

We note that the time measurement taken in the inertial frame is the proper time, and this is t' = 3.0 s.

Then we need to obtain t, for which:

$t = t'/[1 - v^2/c^2]^{1/2}$

where $v = 0.95c$, so:

$t = 3.0 \text{ s}/[1 - (0.95c)^2/c^2]^{1/2} = (3.0\text{s}) 1/[0.0975]^{1/2}$

and $t = (3.0 \text{ s})(3.2) = 9.6\text{s}$

Other Problems:

1) With what speed would a clock have to be moving to run at a rate that is one half the rate of a clock at rest?

2) An atomic clock is placed on a Jumbo Jet. The clock measures a time interval of 3600 s when the jet is moving at $v = 300$ m/s. What corresponding time would an identical clock left on the ground measure? (Hint: whenever $v \ll c$ (e.g. $v/c \ll 1$), note that we have $1 + v^2/2c^2$ and not $[1 - v^2/c^2]^{1/2}$)

3) A muon formed high in the Earth's atmosphere travels at $v = 0.99c$ for a distance of 4.6 km before it decays into an electron, a neutrino and an anti-neutrino.

a) How long does the muon survive as measured in its rest frame?

b) How far does the muon travel as measured in its frame?

4) The average lifetime of a pi meson in its own frame

of reference is 2.6 x 10⁻⁸ s. If the meson moves with v = 0.95c, what is its mean lifetime as measured by an observer on Earth?

Solutions:

1) We have: $t = t'/ [1 - v^2/c^2]^{1/2}$

But the proper time is defined such that:

$t' = t/2$ or $t'/t = 1/2$

Then:

$[1 - v^2/c^2] = (t'/t)^2$

and:

$v^2/c^2 = 1 - (t'/t)^2$

$v^2 = c^2[1 - (t'/t)^2]$

so:

$v = c[1 - (t'/t)^2]^{1/2} = c[1 - 0.5^2]^{1/2} = c[0.75]^{1/2} = 0.866c$

2) The proper time $t' = 3600$ s

Since $v = 300$ m/s $= (10^{-6})$ c and hence v/c << 1 we need the form: $t = t'/ [1 + v^2/2c^2]$ t = 3600s/ [1 + $(10^{-12})c^2/2c^2$] Since the numerator is only slightly larger than 1, the time t will be:

3600 s/(1.000000000001)= 3600.0000000018

= 3600 + 1.8 x 10^{-9} s or slightly longer than one hour.

The proper time t' applies to the muon's reference frame.

So: $t = t'/ [1 - v^2/c^2]^{1/2}$ and $t' = t [1 - v^2/c^2]^{1/2}$

where $v = 0.99 c$ and $v^2 = (0.99c)^2 = 0.98c^2$

Then: $t' = t [1 - 0.98c^2/c^2]^{1/2} = t [0.02]^{1/2} = t(0.14)$

recall distance travelled = 4.6 km = 4600 m

To get t' we need to find t first, e.g.

t = 4600m/ (2.97 x 10^8 m/s) = 1.55 x 10^{-5} s

Then: t' = (1.55 x 10^{-5} s) (0.14) = 2.1 x 10^{-6} s

b) The distance traveled in its frame is just the proper length, L' so:

L' = 4600 m $[1 - v^2/c^2]^{1/2}$ = 4600m $(0.02)^{1/2}$

L' = 4600 m (0.14) = 644 m

4) The proper time t' = 2.6 x 10^{-8} s

$t = t'/ [1 - v^2/c^2]^{1/2}$ and v = 0.95c

so:

t = (2.6 x 10^{-8} s)/ $[1 - (0.95c)^2/c^2]^{1/2}$

$t = (2.6 \times 10^{-8} \text{ s})/ [0.0975]^{1/2} = (2.6 \times 10^{-8} \text{ s})/ 0.312$

$t = 8.3 \times 10^{-8} \text{ s}$

We now extend the special relativity dynamic to place our two observers ("Bill" and "Bob") in the roles of two astronauts in spaceships referred to systems S and S'. Initially each is moving at a constant speed relative to the other. Subsequently, Bob is accelerated to some higher speed so that he is displaced. To return to Bill, Bob will have to slow down at some point (decelerate) then speed up again to overtake Bill, and finally change his speed once more to fall in step with Bill.

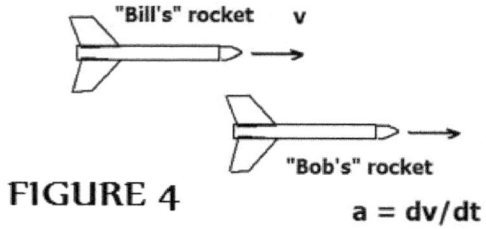

FIGURE 4

During this whole time interval, Bill hasn't experienced any acceleration, but has maintained his constant normal speed. We now ask what effect the acceleration would have had on Bob? It turns out that time has passed more slowly for him and so he's aged less than Bill. If twins to begin with, Bill would now be older than Bob.

Superficially, at least, one may be tempted to dispose of this improbable conclusion using a simple symmetry argument, i.e. basing our coordinate system on Bill's reference frame. This is *wrong* because there is no fundamental difference between the twins: one of them (Bob) has undergone acceleration (experienced a force) which the other (Bill) *has not*.

While all positions and speeds are equivalent in relativity theory, accelerations do have physical consequences, so the relationship between the two twins is not symmetric in this instance and there can be no symmetry argument precluding aging of the un-accelerated twin.

This result of special relativity (which some diehards still can't accept) is the basis for lots of scifi stories that depict space explorers roaming the galaxy at great speeds and returning to Earth after having aged only a few years while millennia have passed on the home planet in the meantime. This effect is called **time dilation** and there is an experimental basis for it.

It has been found that mesons moving very rapidly have a longer average lifetime than mesons not in motion. In fact, this very difference was the basis for two problems given previously. To recap just one of the problems (#4):

The average lifetime of a pi meson in its own frame of reference is 2.6×10^{-8} s. If the meson moves with $v = 0.95c$, what is its mean lifetime as measured by an observer on Earth?

We found the proper time t' = 2.6×10^{-8} s for the meson's own frame

and for the mean meson lifetime for an Earth observer:

$t = t'/ [1 - v^2/c^2]^{1/2}$

$t = (2.6 \times 10^{-8} \text{ s})/ [1 - (0.95c)^2/c^2]^{1/2}$

$t = (2.6 \times 10^{-8} \text{ s})/ [0.0975]^{1/2} = (2.6 \times 10^{-8} \text{ s})/ 0.312$

$t = 8.3 \times 10^{-8}$ s

In other words, t > 3t' or, more than three times as much time has elapsed for the Earth observer in his stationary frame.

Readers can check a similar sort of result was found for the muon moving at relativistic speeds in problem (3).

The dilation of time has even been found to occur when comparing the clocks in a Jumbo Jet and those on the ground. This was the basis for problem (2) in the last set which found that the clock time aboard a Jumbo Jet traveling at 300 m/s indicated an hour clocked on Earth as $3600 + 1.8 \times 10^{-9}$ s, or slightly longer than one hour (by about two billionths of a second). Such a tiny difference - which is expected for relatively low velocities, e.g. much lower than c, can be determined by a high precision atomic clock.

FIGURE 5

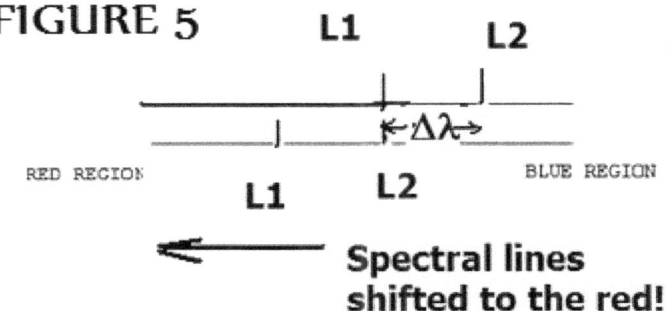

NOTE: $\Delta\lambda/\lambda = v/c$

An additional example is provided in the well-known Doppler effect (Fig. 5) by which any form of electromagnetic radiation changes its wavelength and frequency as the source moves toward or away from an observer. The Doppler formula for z, the red shift is: $z = (\Delta\lambda/\lambda) = v/c$

tells us exactly how this change should vary with velocity. However, when a source is moving very close to light speed we will find the change doesn't quite agree with what is observed, so the formula must be modified for relativistic effects - just as ordinary motions must be modified whenever $v \sim c$. Thus, the correct relativistic effect (say for changing frequency) is given by:

$f_o/f = [[1 - v^2/c^2]^{1/2} / (1 + v/c) =$

$[(1 - v/c)^{1/2}]/(1 + v/c)^{1/2}]$

In terms of z, the red shift

$1 + z = [1 + v/c]^{1/2}/[1 - v/c]^{1/2}$

Note that z tends to infinity when v approaches c.

In any problems where z becomes comparable to 1, one must apply the relativistic formula. Failure to do so results in an incorrect interpretation that the redshifts indicate recessional velocities greater than light speed.

Problems:

1) Assume two astronauts are traveling at v = 0.95c on a journey to the system of Alpha Centauri. We on Earth would say that it takes 4.2 / 0.95c = 4.4 years to reach the system 4.2 light years distant. But the astronauts dispute this.

(a) How much time passes on the astronauts' clocks?

(b) What is the distance to Alpha Centauri as measured by the astronauts? (Hint: this is an exact analog of the muon path length problem (#3) from the previous problem set)

2) According to Hubble's law, the distant galaxies are receding from us at speeds proportional to their distances, d, e.g. v = Hd. (Where $H = 2.26 \times 10^{-18}$ s^{-1}, currently).

a) How far away would a galaxy be in light years whose velocity relative to the Earth is c?

b) Would it be observable from Earth? (Take 9.5×10^{15} m = 1 LY)

3) A galaxy in Hydra emits light with a red shift corresponding to a recessional velocity of 6×10^4 km/s.

a) What is its distance according to Hubble's law?

b) What is the value of z?

c) Assume this galaxy passed Earth T years ago and has moved with constant velocity ever since, what is the value of T?

4) Some observations reported on the quasar 3C-9 suggest that when it emitted the light that just reached Earth it was receding at a velocity of 0.8c. One of the lines identified in its spectrum has a wavelength of 1200 Å (angstroms) when emitted from a *stationary source*.

a) At what wavelength must this spectral line have appeared in the observed spectrum of the quasar?

b) What is its red shift, z?

Solutions:

1) Let t_A be the time on the astronauts' clock and t_E be the time recorded on an Earth-based clock.

Then, we have $t_E = 4.4$ yrs.

And:

$t_A = t_E [1 - v^2/c^2]^{1/2}$

$t_A = (4.4 \text{ yrs.}) [1 - (0.95c)^2/c^2]^{1/2} = 1.37$ yrs.

(b) Since we know: $t_A = 1.37$ yrs.

then the distance $D_A = (0.95c)(1.37 \text{ yrs}) = 1.31$ Ly

2) In this case, $v = c = 3 \times 10^8$ m/s

$d = v/H = (3 \times 10^8 \text{ m/s})/(2.26 \times 10^{-18} \text{ s}^{-1})$

$d = 1.32 \times 10^{26}$ m

Converting to light years:

$d = (1.32 \times 10^{26} \text{ m})/(9.5 \times 10^{15} \text{ m /Ly}) = 1.4 \times 10^{10}$ Ly

b) Would it be observable from Earth?

Given that modern telescopes can penetrate to about 1.8×10^{10} Ly, the galaxy should easily be observable to the Hubble but might be more problematical for land-based scopes.

3) We know the recessional velocity $v = 6 \times 10^4$ km/s

By Hubble's law: $v = Hd$ so the distance $d = v/H$

Then, attending to the proper units for v, H:

d = (6 x 10^7 m/s)/(2.26 x 10^{-18} s^{-1})= 2.6 x 10^{25} m

and d = (2.6 x 10^{25} m)/(9.5 x 10^{15} m /Ly) = 2.8 x 10^9 Ly

(b) z = v/c = (6 x 10^7 m/s)/(3 x 10^8 m/s) = 0.2

(c) T = d/v = (2.6 x 10^{25} m)/(6 x 10^7 m/s) = 4.3 x 10^{17} s

But 1 yr. = 3.15 x 10^7 s

so T = (4.3 x 10^{17} s)/(3.15 x 10^7 s/ yr)

T = 1.36 x 10^{10} years, or 13.6 billion years

4) Let λ_o be the normal wavelength = 1200 Å and λ be *the red-shifted* value.

We know v = 0.8c so we must use the modified Doppler version, viz.

λ/λ_o = $(1 - v/c)^{1/2}$ /$(1 + v/c)^{1/2}$

λ/λ_o = $(1 + 0.8)^{1/2}$/ $(1 - 0.8)^{1/2}$ = $(1.8/0.2)^{1/2}$

λ/λ_o = $\sqrt{9}$ = 3

then:

λ = 3 λ_o = 3 (1200 Å) = 3600 Å

(b) The red shift of the quasar is found from:

$1 + z = (1 + v/c)^{1/2} / (1 - v/c)^{1/2}$

$1 + z = (1.8/0.2)^{1/2} = \sqrt{9} = 3$

Then: $z = 3 - 1 = 2$

Other Problems:

1) Imagine a space shuttle service from Earth to Mars, and each ship is equipped with 2 identical lights – one at the front and one at the rear. The two ships normally travel at a velocity v_o relative to the Earth such that the head light of a ship approaching Earth appears green (e.g. $\lambda = 5000$ Å) and the tail light of a departing ship appears red ($\lambda = 6000$ Å).

From this, find the ratio: v_o/c.

2) Given the above problem info, assume one ship accelerates to overtake the other (ahead of it). At what velocity relative to Earth must the overtaking ship travel so the tail light of the ship ahead of it looks like a head light, e.g. green or 5000 Å?

3) A Calcium line in the spectrum of α Centauri has a wavelength of 3968.20 Å. The same line in the solar spectrum has a measured wavelength of 3968.49 Å.

Find the radial velocity of α Centauri relative to the solar system. Is it approaching or receding?

XXII. Relativity (II): The Lorentz Contraction

Having dealt with time in the context of special relativity, we now consider what happens to length. We return to the diagram of systems S and S' moving relative to each other (Fig. 1) and consider a meter stick of length L (= 1 m) pointed in the +x direction and moving in that direction with velocity v.

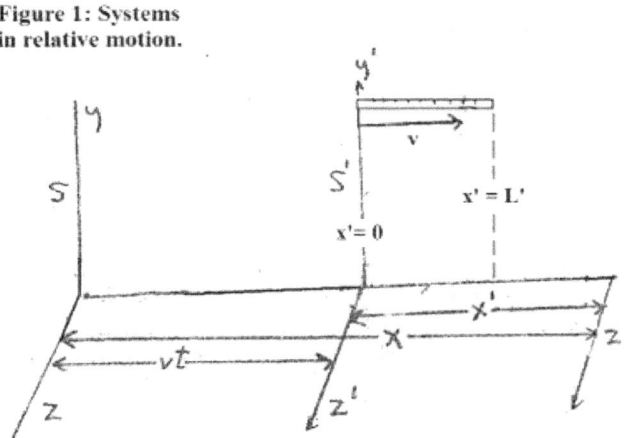

Figure 1: Systems in relative motion.

We can also think of it as being at rest in the S' system with one end at x' = 0 and another end at x' = L', initially.

Its length in the system S' is therefore L'. At time t = 0 we take a photograph of the meter stick with a camera located in the S system. We inquire: What are the positions of the two ends of the meter stick in the S system? We can answer this by using the first of the Lorentz-Einstein transformations:

$x = x' + vt'/(1 - v^2/c^2)^{1/2}$

and

$x' = x - vt/ (1 - - v^2/c^2)^{1/2}$

We opt to use the lower form rather than the upper, having already specified an instant at t = 0 and hence looking for simplification. Then:

$x' = x / (1 - v^2/c^2)^{1/2}$

Now, the end of the meter stick that is at x' = 0 in S' is found to be at x = 0 in S. The *other* end in S is seen to be at:

$x = L' (1 - v^2/c^2)^{1/2}$

This informs us the length of the moving meter stick in S so we can say that the length of the moving meter stick as seen by the camera is:

$L = L' (1 - v^2/c^2)^{1/2}$

As can readily be seen, this implies *length contraction*. For example, say the rod is moving at v = 0.6c, then the length L is (given L' = 1m):

$L = (1 \text{ m}) (1 - (0.6c)^2/c^2)^{1/2} = 1m [(1 - 0.36c^2/c^2)^{1/2}$

$L = (1m) (0.64)^{1/2} = 1m (0.8) = 0.8 \text{ m (or 80 cm)}$

(Remember that by symmetry arguments of relativity, the observer in the other system will argue that L' = L

$(1 -v^2/c^2)^{½}$ so that *from his viewpoint* a meter stick will be similarly foreshortened.)

Readers may recall that something similar occurred in the time transformation, i.e. each of two relatively moving clocks ran slower with respect to the other than to itself.

This brings us back to the solution of the Michelson-Morley experiment discussed in an earlier chapter. It is the curious, symmetric relativity in time and length (in direction of motion) which solves the paradox of the "missing ether wind" quite simply and logically. Moreover it is solved more fundamentally and satisfactorily than either Fitzgerald or Lorentz could manage.

Let's derive the new law for addition of velocities. Assume an object in the S' system starts at the point x' = 0 at time t' = 0. It moves with constant velocity u' (relative to S') and in the time t' it travels a distance x'.

By definition, $u' = x'/t'$

We ask: 'How fast does this object travel according to the observer at rest in System S?'

This observer will see a velocity given by the formula:

$u = x/t = (x' + vt')'/ t' + x'v/c^2$ or

$u = (t' + v)/(1 + x'v/t' c^2)$

$u = (u' + v)/ 1 + u'v/ c^2$

This is *the relativistic formula for addition of velocities*. If u' and v = c then the formula yields:

$u = (c + c)/ 1 + c^2/c^2$

$u = 2c/ 1 + 1 = 2c/2 = c$

If u' and v *are less than* c then u must always be less than c.

Example Problem:

Say two objects are moving at 3c/4 towards each other, then what is their relative velocity as recorded in a system S observing their approach?

We have: u' = 3c/4 and v = 3c/4

then:

$u = (u' + v)/ 1 + u'v/c^2 = (3c/4 + 3c/4)/1 + u'v/c^2$

where: $(3c/4 + 3c/4) = 3c/2$

and:

$u' v = (3c/4)(3c/4) = 9c^2/16$

Then:

$u = (3c/2)/ [1 + 9c^2/16/c^2]$

$u = (3c/2)/ (1 + 9/16) = (3c/2)/ (25/16)$

$u = (3c/2)(16/25) = 24c/25$

Other Problems:

1) Assume that the lifetime of quasar 3c-9 is 1 million years as measured in its own rest frame. Over what total time span (in Earth-measured time) would its radiation be received at the Earth? (Assume 3c-9's velocity relative to the Earth remains constant)

2) Suppose that you happen to be moving at a velocity of 3c/4 past a remote observer who picks up a stopwatch and then sets it down. Using a high power telescope you observe he held the watch for 9 seconds. How long would HE think that he held it?

3) An astronaut orbits the Earth at a distance of 7 x 10^6 m from its center for a week. How much younger than his twin on Earth is he when he lands? Assume standard orbital speed of 18,000 mph and neglect the rotation of the Earth.

4) Consider 3 galaxies: A, B and C. An observer in A measures the velocities of B and C and finds they are moving in opposite directions - each with a speed of 0.7c relative to him., i.e.

(0.7c) <-----------*(B)*-----*(A)*-----*(C)*--------->*(0.7c)*

What is the speed of A observed by someone in B?

What is the speed of C observed by someone in B?

The observer in A thinks that the two other galaxies are receding from him at a rate 1.4c. Show him how this is wrong, by providing the correct result.

Solutions:

1) Let $t = 10^6$ yrs be the quasar's lifetime in its own rest frame. Then the total Earth-based time that its radiation will be received is:

$t' = t [1 - v^2/c^2]^{1/2}$

where $v = 0.8c$

so:

$t' = (10^6 \text{ yrs.}) [1 - (0.8c)^2/c^2]^{1/2}$

$t' = (10^6 \text{ yrs}) [1 - 0.64]^{1/2} = 10^6 \text{ yrs } (0.36)^{1/2}$

$t' = 0.6 (10^6 \text{ yrs}) = 6 \times 10^5 \text{ yrs.}$

(2) Let $\Delta t' = (t_2' - t_1') = 9 \text{ s}$

$t_2' = (t_2 - 3c/4)(1 - 9/16))^{-1/2}$

$t_1' = (t_1 - 3c/4)(1 - 9/16)^{-1/2}$

or:

$t_2' = (t_2 - 3c/4) (7/16)^{-1/2}$

$t_1' = (t_1 - 3c/4) (7/16)^{-1/2}$

whence:

$(t_2' - t_1') = [(t_2 - 3c/4) - (t_1 - 3c/4)]/0.661$

or: $9 \text{ sec} = (t_2 - t_1)/0.661$ and

$\Delta t = 5.95 \text{ sec}$

(3) The speed of the astronaut is given by:

$v = (2gr)^{1/2}$

and $r = 7 \times 10^6$ m, $g = 10$ ms^{-2}

$v = 11,832$ m/s $= 3.94 \times 10^{-5}$ (c)

In the astronaut's frame, 1 week - 86,400 s x (7) = 604, 800 s

For the twin on Earth:

$t = t'/[1 - t^2/c^2]^{1/2}$

$t = 604,800.0002$s

and so the time difference = 0.0002s, or the twin in orbit will be (t' - t) or 0.0002 s younger on his return.

(4) The speed of A observed in B = 0.7c, exactly equal to the speed of B observed in A, by principle of relative velocities.

To find the speed of C observed in B, we use

relativistic addition of velocities or:

$u = (u' + v)/ 1 + u'v/c^2$

$u = (0.7c + 0.7c)/ [1 + (0.7c)(0.7c)/c^2]$

$u = (1.4c)/ 1 + 0.49 = 1.4c/1.49 = 0.94c$

Other Problems to solve on your own:

1) A rocket ship of length 100m travels at $v/c = 0.6$. It carries a radio receiver in its nose. A radio pulse is emitted from a stationary space station just as the ship passes by.

a) How far from the space station is the nose of the rocket at the instant the radio signal arrives at the nose?

b) By space station time, what is the time interval between the arrival of this signal and its emission from the station?

c) What is the time interval determined from measurements in the rocket ship's rest frame?

2) A flash of light emitted at position x_1 on the x-axis is absorbed at position $x_1 + \ell$. In a reference frame moving with velocity $v = \beta c$ along the axis, what is the spatial separation ℓ' between the point of emission and point of absorption of the light? ($\beta = 1/ [1 - v^2/c^2]^{1/2}$)

How much time elapses between emission and absorption of the light?

XXIII. The Inertia of Energy

We now conclude our introduction to Einstein's theory of special relativity with what many regard as its most fundamental conclusion: the inertia of energy, or light. This is embodied in Einstein's famous equation:

$E = m c^2$

which is more accurately posed as:

$E = (\Delta m) c^2$

where Δm is the "mass defect" or difference, say in a nuclear reaction, and c is the velocity of light.

Before looking at examples, it's useful to consider the relativistic mass of a particle, in terms of its rest mass m_o. The rest mass, as the term implies is the mass of the object at rest or:

$m_o = m [(1 - v^2/c^2)^{1/2}]$

Thus, if $v = 0$ (particle at rest) then we have:

$m_o = m(1)^{1/2} = m$

so the mass and rest mass are identical.

Now, the relativistic mass is then:

$m = m_o / [(1 - v^2/c^2)^{1/2}]$

and again, if $v = 0$ then $m_o = m$

But what if $v = c$? (Object moving at the speed of light?)

Then:

$m = m_o / [(1 - c^2/c^2)^{1/2}] = m_o / [1-1]^{1/2} = m_o / 0 = \infty$

In other words, m would be infinite! This is another way of saying that to try to achieve the velocity of light one would have to overcome infinite inertia! In other words, it can't be done...not for a material object.

From this, we can also see the relativistic momentum must be:

$p = mu = m = m_o u / [(1 - c^2/c^2)^{1/2}]$

This approaches the classical value ($p = mu$) as $u \to 0$

Newton's 2nd law in the relativistic format is simply:

$F = ma = m(du/dt) = d/dt[m_o u / [(1 - v^2/c^2)^{1/2}]]$

The relativistic energy is found by taking the integral of: $(dp/du) u\, du$

$\to \int_0^u u\, dp$

from 0 to u and obtaining:

$W = mc^2 / [(1 - u^2/c^2)^{1/2}] - mc^2$

And by the work-energy theorem:

W = K(f) - K(i)

where K(i) is just the initial rest energy, or $m_o c^2$

Then W = $m_o c^2 / [(1 - u^2/c^2)^{1/2}] - m_o c^2$ =

(total energy - rest energy)

Example Problem:

Apply the basic mass-energy equation, E = $(\Delta m) c^2$, to the case of nuclear fusion.

Consider:

$_1H^2 + {}_1H^2 \rightarrow {}_2He^3 + {}_2He^3 + {}_0n^1$

which actually occurs in the Sun.

We now compile the masses (in atomic mass units) on each side:

2.015 u + 2.015 u → 3.017 u + 1.009 u

or:

4.030 u → 4.026 u

Now, *the right side is less than the left* by an amount equal to the mass defect or:

$\Delta m = 4.030\ u - 4.026\ u = 0.004\ u$

To get the energy E:

$E = (0.004\ u)(931\ MeV/u) = 3.7\ MeV$

where 931 MeV/u is the conversion factor incorporating c^2

To transfer to Joules:

$3.7\ MeV = 3.7\ MeV \times (1.6 \times 10^{-13}\ J/MeV) = 6.0 \times 10^{-13}\ J$

Example Problem (2):

Determine the energy required to accelerate an electron from 0.50c to 0.90c.

By the work-energy theorem:

$W = K(f) - K(i)$

$K(i) = m_o c^2 / [(1 - u^2/c^2)^{1/2}]$

$u_1 = 0.50\ c$

$K(f) = m_o c^2 / [(1 - u^2/c^2)^{1/2}]$

$u_2 = 0.90c$ (where: $m_o = 9.1 \times 10^{-31}$ kg)

$K(f) - K(i) = m_o c^2 / [(1 - (0.90c)^2/c^2)^{1/2}]$

$- m_o c^2 / [(1 - (0.50c)^2/c^2)^{1/2}]$

$K(f) - K(i) = m_o c^2/[(1 - 0.81)]^{1/2} - m_o c^2/[(1 - 0.25)]^{1/2}$

$K(f) - K(i) = 2.294 \, m_o c^2 - 1.155 \, m_o c^2 = 1.134 \, m_o c^2$

Or:

$K(f) - K(i) = 9.32 \times 10^{-14}$ J $= 0.583$ MeV

Other Problems:

(1) A spaceship of mass 10^8 kg is to be accelerated to 0.6c using a matter-antimatter mix engine.

(a) How much energy does this require?

(b) How many kilograms of matter and antimatter will it take to provide this much energy?

(2) Consider the decay:

$_{24}Cr^{55} \rightarrow {_{25}}Mn^{55} + e^-$

The Cr 55 nucleus has a mass of 54.9279 u and the Mn 55 nucleus has a mass of 54.9244u.

(a) Calculate the mass difference between the two nuclei.

(b) What is the maximum kinetic energy of the emitted electrons?

(3) Find the energy required to remove a simple proton from $_{19}K^{41}$.

(4) Find the speed and mass of an electron whose kinetic energy is 50 MeV.

(5) A rocket ship is to be accelerated to a speed of 0.5c. If propulsion is to be by using nuclear fuel, what fraction of the initial rest mass of the ship would have to be converted into kinetic energy to attain the desired speed?

What time dilation results if the speed is v = 0.5c?

Would this be sufficient to allow one generation of humans to reach the star Proxima Centauri (4.2 light years distant)?

Solutions:

1.(a) by *the work -energy theorem*:

$W = K(f) - K(i)$

or:

$W = m_o c^2 / [(1 - u^2/c^2)^{1/2}] - m_o c^2$

where $m_o = 10^8$ kg and u = 0.6c

Then:

$W = (10^8 \text{ kg})c^2 / [(1 - (0.6c)^2/c^2)^{1/2}]$

$W = (10^8 \text{ kg})c^2 / ([1 - 0.36]^{1/2}) = (10^8 \text{ kg})c^2 / ([0.64]^{1/2})$

$W = (10^8 \text{ kg})c^2 / (0.8) = 1.25 \times 10^8 \text{ kg}(c^2) = 1.12 \times 10^{25}$ J

(b) According to theory, equal amounts (masses) of matter and antimatter are required for complete annihilation and total conversion of the initial masses into energy. The energy needed (as shown in (a)) is 1.12×10^{25} J. Thus: $m(a) = m(m) = 6.25 \times 10^7$ kg.

(2) The Cr 55 nucleus has a mass of 54.9279 u and the Mn 55 nucleus has a mass of 54.9244u, hence, the mass difference is:

$\Delta m = 54.9279\ u - 54.9244u = 0.0035\ u$

(b) The maximum kinetic energy of the emitted electrons can be found using the mass defect.

The mass defect (from (a)) is 0.0035u so:

$E = (\Delta m)\ c^2 = (0.0035u)(931\ MeV/u) = 3.25\ MeV$

By the work-energy theorem:

$W = K(f) - K(i)$

$K(i) = 0\ (u_1 = 0)$

$K(f) = m_o\ c^2 / [(1 - u^2/c^2)^{1/2}] = 3.25\ MeV$

where we need to find u.

Using a table one finds the rest energy of the electron = 0.511 MeV

and $K(f) = (3.25/ 0.511) = 6.36x$ the rest mass

So:

$E = 6.36mc^2 = mc^2 / [(1 - u^2/c^2)^{1/2}]$

$6.36 = 1/[(1 - u^2/c^2)^{1/2}]$ or:

$(1 - u^2/c^2) = 1/(6.36)^2 = 1/40.44$

or:

$u^2/c^2 = 1 - 0.0024 = 0.9976$

Therefore: $u = [0.9976c^2]^{1/2} = 0.9987c$

(3) In removing a single proton from $_{19}K^{41}$ the atom remaining is: $_{18}A^{40}$ (e.g. $_{19}K^{41} \rightarrow {}_{18}A^{40} + p+$)

whose measured mass is: 39.96238 u

The mass of the final system becomes:

39.96238 u + 1.007825 u = 40.970205 u

The effective *mass increase* of the system:

Δm = 40.970205 u - 40.96184 u = 0.008365 u

Then the energy needed to remove a single proton is, from Einstein's eqn.:

$E = (\Delta m) c^2$ = (0.008365 u) (931 MeV/u) = 7.79 MeV

(Which can also be converted to MJ using: 1 MJ

= 1.6 x 10⁻¹³ MJ/MeV)

Thus: E = (7.79 MeV)(1.6 x 10⁻¹³ MJ/MeV) =

1.24 x 10⁻¹² MJ

(4) As per Problem #2, we know the electron rest mass = 0.511 MeV

Then, the mass of this electron is:

(50 MeV/0.511 MeV) = 97.84

or 97.84 *times the normal electron rest mass.*

The speed, following the same procedure applied in #2, is obtained by first solving for the energy, E:

$E = 97.84 \, mc^2 = mc^2 / [(1 - u^2/c^2)^{1/2}]$

Then we need u:

$(1 - u^2/c^2) = 1/(97.84)^2 = 1/9572$

or: $u^2/c^2 = 1 - 0.0001 = 0.9999$

$u = [0.9999c^2]^{1/2} = 0.9999c$ (approx.)

(5) By the work -energy theorem:

W = K(f) - K(i)

where K(i) is just the initial rest energy, or $m_o c^2$

Then $W = mc^2/[(1 - u^2/c^2)^{1/2}] - m_o c^2 =$

(total energy - rest energy)

Then, the total energy E:

$= mc^2/[(1 - u^2/c^2)^{1/2})^{1/2}] = mc^2/[(1 - 0.25)^{1/2}] = 1.16mc^2$

Then *the excess* $(E - E' = E_k)$ is that required from the fuel, or:

$E_k = (1.16mc^2 - mc^2 = 0.16mc^2)$

Mass of nuclear fuel - call it m'- is then related to E_k by:

$E_k = m'c^2 = 0.16m'c^2$ or

$(m'/m) = 0.16$ or $m' = 0.16m$

In other words, the nuclear fuel mass(m') needs to be at least 16% of the total initial mass of the rocket. So, if the rocket's mass is 100,000 kg then the nuclear fuel must be at least (0.16)(100,000 kg) = 16, 000 kg.

For time dilation:

$t' = t[1 - v^2/c^2]^{1/2}$

again, v = 0.5 so:

$t' = t[1 - 0.25]^{1/2} = t[0.75]^{1/2} = 0.866t$

Now, there are 9.5×10^{15} meters per light year

Therefore the distance to Proxima Centauri is:

D = (4.2 Ly) x (9.5 x 10^{15} m/Ly) = 4 x 10^{16} m

Or: D = 4 x 10^{13} km

The time required is:

(0.866/0.500) x (4 x 10^{13} km)/ (300,000 km/s]

= 1.73 x (4.22 yrs) = 7.3 years

Since a generation is generally figured as 40 years, the time factor will not be an issue

Extra Relativity problems to solve:

1) Observations on the quasar 3C-9 indicate that when it emitted the light that just reached Earth it was receding at a velocity of 0.8c. One of the lines identified in its spectrum has a wavelength of 1200 Å when emitted from a stationary source. At what wavelength must this line have appeared in the spectrum of 3C-9?

2) Assume the lifetime of quasar 3C-9 is 10^6 years when measured in its own rest frame. Over what total span of Earth time would its radiation be received at the Earth?

3) An astronaut circles the Earth at a distance of 7600 km (from its center) for a week. How much younger than a twin remaining on Earth is he when he lands.

OPTICS SUPPLEMENT

1. Refraction

Given the critical importance of optical instruments, including refractors and reflecting telescopes to astronomy, we now begin a survey of light and optics. We start with one of the most fundamental investigations: that of refraction. In particular, we are interested in confirming *Snell's Law*, depicted in the diagram below:

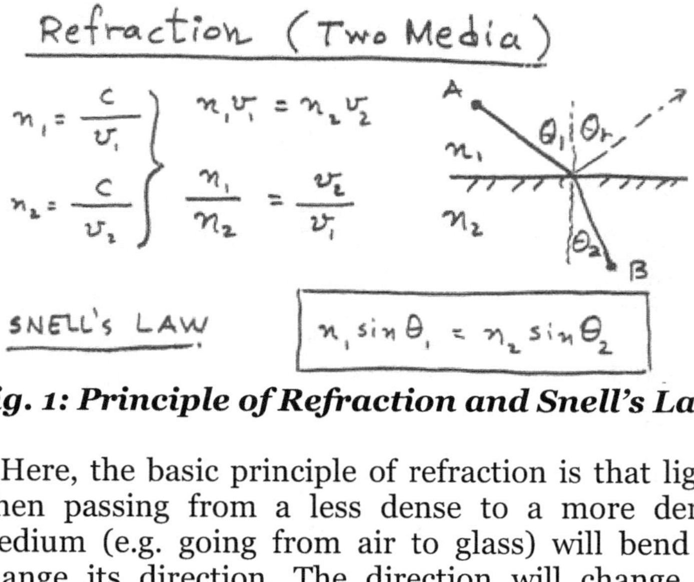

Refraction (Two Media)

$$n_1 = \frac{c}{v_1}$$
$$n_2 = \frac{c}{v_2}$$
$$n_1 v_1 = n_2 v_2$$
$$\frac{n_1}{n_2} = \frac{v_2}{v_1}$$

SNELL'S LAW: $n_1 \sin \theta_1 = n_2 \sin \theta_2$

Fig. 1: Principle of Refraction and Snell's Law

Here, the basic principle of refraction is that light, when passing from a less dense to a more dense medium (e.g. going from air to glass) will bend or change its direction. The direction will change on entering, then change again on exiting - since the latter case reverses the media (from more dense to less dense).

In the diagram n_1 and n_2 refer to refractive indices, defined as:

$n_1 = c/v_1$ and $n_2 = c/v_2$

thus, n_1 is the refractive index *for air*, and n_2 the index for glass. If we require:

$n_1 v_1 = n_2 v_2$, or $n_1/n_2 = v_2/v_1$

In other words, the ratio of the refractive index in the denser medium to the refractive index in the less dense medium is the ratio of the velocity of light in the denser medium to the value in the less dense medium. Taking into account the two angles: Θ_1 = angle of incidence, and Θ_2 = angle of refraction, we may write:

$n_1 \sin \Theta_1 = n_2 \sin \Theta_2$

which is Snell's law. The experimental set up is shown below in Fig. 2 :

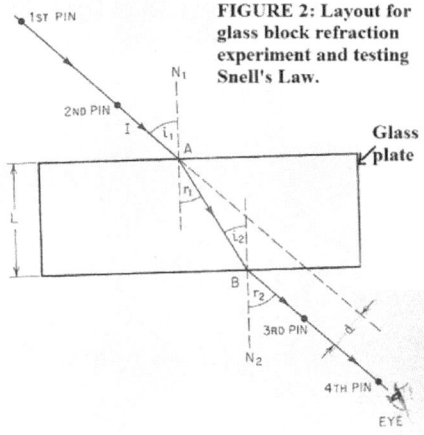

FIGURE 2: Layout for glass block refraction experiment and testing Snell's Law.

Here we set a glass plate or rectangular block on a sheet of A4 paper (toward the center) then use two optical pins to define the incident ray with respect to the normal (N) to the top of the block. (The normal line is carefully extended into the paper to more easily construct the rays.) We then use a 3rd and 4th pin to identify the exit ray from the lower portion of the glass block, and reference the angle r2 carefully by line of sight. Once the incident ray and emergent ray are validated, it is possible to complete the other measurements, for i1 and r1 and then for i2 and r2. By Snell's law then:

sin i/ sin r = v1/v2 = n

and it should be possible to obtain the index of refraction of the glass block.

Example problem:

In the experiment set up from Fig. 2, a student measures angle i1 = 60 degrees and r1 = 30 degrees. Find the experimental value of n for the block.

Solution:

We have:

sin i1/ sin r1 = n

so: sin (60)/ sin (30) = n

or n = (√3/ 2)/ ½ = √3 = 1.73

Other *problems*:

1) For the same set up and experiment, obtain the values the student should get for the angles i2 and r2

for the emergent ray if his results are to be consistent.

2) If the speed of light in air is 300,000 km/sec what is the speed in a substance with refractive index 1.74?

3) Light in air has a wavelength of 0.0000589 cm. What would the wavelength be in water for which the refractive index for water to air is $n_w = 4/3$.

4) When the angle i1 is equal to $\pi/2$, i2 is equal to what is called the **critical angle**. (For this case note: $\sin(\pi/2) = 1$ so: $1 = n \sin(i2)$.) Find the critical angle here.

5) If the index of refraction of a piece of extra dense flint glass is 1.60, would the critical angle be greater or less than 60 degrees?

2. Thin Lenses (Convex)

The investigation of thin lenses is of practical import with applications to telescopes, binocular, cameras and projectors as well as eye glasses.

Experimental set up: To Find Lens Constants:

Preliminary Note:

An optical bench is recommended for these experiments but not critical. A meter rule can also be used with clay holders for lenses and a small flashlight

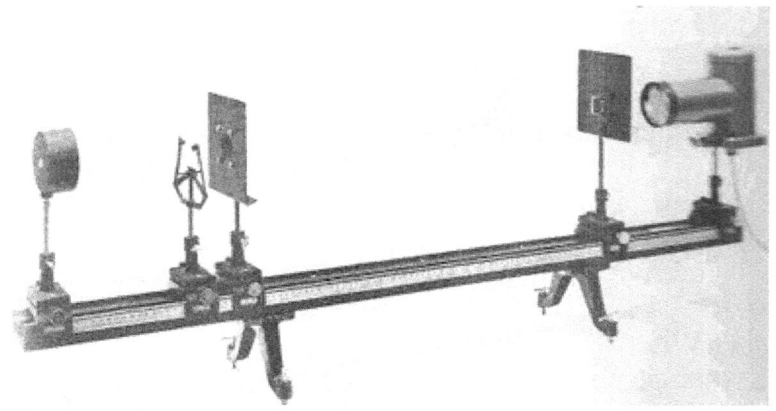

Fig. 3: Optical bench with lenses, objects in place to measure focal length of a thin lens.

A standard setup is shown in Fig. 3 above, for which an optical bench is used. The bench includes at least three sliding mounts, at least three lenses of differing focal lengths (e.g. 20 - 30 cm), an illuminated wire screen to serve as an object, a hooded screen, and an image screen set up in conjunction with a magnifier.

The theoretical results are depicted in the graphic (Fig. 1(b)) below, with the Object O positioned on the left side of the lens, which has a radius of curvature R2, and center of curvature C2, and an image I on the right side of the lens.

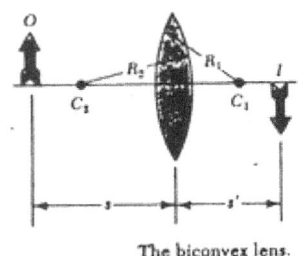

The biconvex lens.

Fig. 4: Image formation with biconvex lens

The lens has a radius of curvature R1 and center of curvature C1. The object distance is then s, and the image distance is s'. These can then be used to form the thin lens formula (top) and the "*lens makers equation*" bottom.

Thin lens formula

$$\frac{1}{s} + \frac{1}{s'} = (n-1)\left(\frac{1}{R_1} - \frac{1}{R_2}\right)$$

paraxial rays
small thickness

Lens makers' equation

focal length f
for $s \to \infty$, $f = s'$

$$\frac{1}{f} = (n-1)\left(\frac{1}{R_1} - \frac{1}{R_2}\right)$$

Thin lens formula and lensmaker's equation.

Procedure:

The bench is set up as shown (Fig. 3) with 4 carrier slides or mounts, one for the object, one for the lens, one for the hooded screen and one for the light source. The lenses chosen for examination are all double convex (thus each will have R1, R2) and be made of ordinary crown glass.

Place the object screen near one end of the bench and the lens near the center, and take measurements: s and s'. Repeat then for a different value of s' such that s < s'. Take spherometer measurements of the lens, to find R1, and R2.

Calculations:

From the initial data s and s' the focal length f can be

found (sing the thin lens equation). Using the values obtained for R1 and R2, the value for n, the refractive index, can be found.

Sample problem:

In the experimental set up shown in Fig. 1, a student obtains the following measurements:

$R_1 = 10$ cm

$R_2 = -4$ cm

$s = 30$ cm

a) If the student is given the refractive index of the lens as $n = 1.52$ how can he find the focal length?

b) Given the above result and the measurement made for s, how can the student find the image distance, s'?

Solution:

a) By the lens maker's eqn.

$1/f = (n - 1) [1/R_1 - 1/R_2]$

then: $f = R_1 R_2 / (R_2 - R_1) \times (1/n - 1)$

$= [(10 \text{ cm})(-4 \text{cm})] / [-4 \text{ cm} - 10 \text{ cm}] \times 1/0.520$

$f = (-40) \text{ cm} / -14 \times 1/0.520 = 5.49$ cm

b) Re-arrange the thin lens eqn. to make s' the subject:

$1/s' = 1/f - 1/s$ or:

$1/s' = s - f / fs$

then:

$s' = fs / (s - f) = (5.49 \text{ cm})(30 \text{ cm}) / (30 \text{ cm} - 5.49 \text{ cm})$

$s' = 6.72 \text{ cm}$

Other problems:

1) Using the experimental information and the lens diagram graphic as a guide, construct a lens diagram and thereby find the height of the image formed in the student's experiment.

2) By a similar graphical construction, find the position of the image formed by a converging lens (f = 15 cm) when:

a) the object is 60 cm from the lens

b) the object is 30 cm from the lens

3) A lamp and a screen are 160 cm apart. What should the focal length of the lens be in order to produce a real image of the lamp three times as large as the source?

(Hint: we can use the proportion: $h'/h = s'/s$ to find

image or object dimensions using image and object distances)

3. Thin Lenses (Concave)

We now conclude our examination of single refracting surfaces as thin lenses, with a look at the diverging lens. In the previous section we focused on converging lens, namely the biconvex lens, and now we focus on the biconcave lens.

FIG. 5

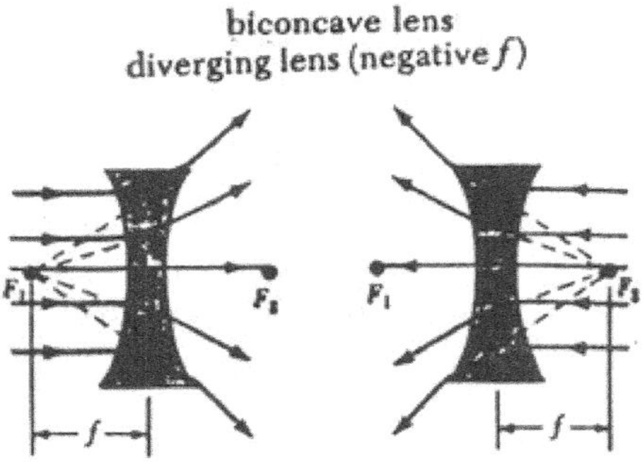

biconcave lens
diverging lens (negative f)

Sign Convention for Thin Lenses

s is $+$ if the object is in front of the lens.
s is $-$ if the object is in back of the lens.

s' is $+$ if the image is in back of the lens.
s' is $-$ if the image is in front of the lens.

R_1 and R_2 are $+$ if the center of curvature is in back of the lens.
R_1 and R_2 are $-$ if the center of curvature is in front of the lens.

Figure 5 shows the basic directions of the diverging rays from foci (F1, F2) of a single biconcave lens. The key point embodied in the diagram is that the focal length (f) is negative. This must be factored into all lens computations (such as when one is asked to find the image and-or object distances) for such lenses.

The bottom of Fig. 5 summarizes all the sign

conventions for use with both converging and diverging thin lenses. It shows how the quantities s, s', R1 and R2 all change their sign in respect of where object or image is placed in reference to the lens, or where the centers of curvature are placed.

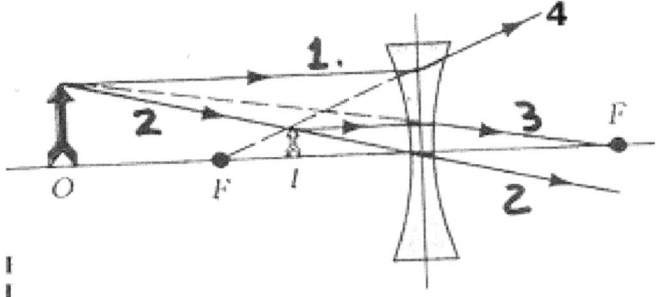

FIG. 6: Ray DIAGRAM for DIVERGING LENS.

Figure 6 shows a typical ray construction as it would be applied to a diverging lens and in many ways is similar to the converging lens except that now the object is formed **on the same side of the lens as the image** and is *larger than the image.* To summarize the numbered rays:

(1) Drawn from the top of the object (O) to the optical axis .

(2) Drawn from the top of O through the center of the lens and then continued as a straight line

(3) Is drawn from the top of I (image) to the lens optical axis, thence to the opposite focus (F)

(4) Is drawn starting from the first focus (F) on the left side of the lens, intersecting the top of I then on toward the optical axis at upper part of lens.

We designate (1)- (3) as *Principal Rays* and (4) as ancillary.

Problems:

(1) The focal length of a diverging lens is 30 cm, and the object is 40 cm away. Find the image distance.

Solution:

We have: f = - 30 cm

s = 40 cm

The thin lens equation for diverging lenses is:

1/s + 1/ s' = - 1/f

and in this case, we will write:

1/40 + 1/ s' = - 1/30

Then:

1/s' = -1/f -1/s = - 1/30 - 1/40

1/s' = - 7/120

s' = -120/7 cm = -17.1 cm

According to our sign conventions (Fig. 1, bottom) s' is negative if the image is in front of the lens. Since s' = - 17.1 cm it must be located 17.1 cm in front of the lens conforming with what we see in the ray diagram of Fig. 2

(2) Using a graphical construction find the image formed by a diverging lens with a focal length of 30 cm when the object is 60 cm from the lens.

Solution: The diagram shown below provides the scale solution.

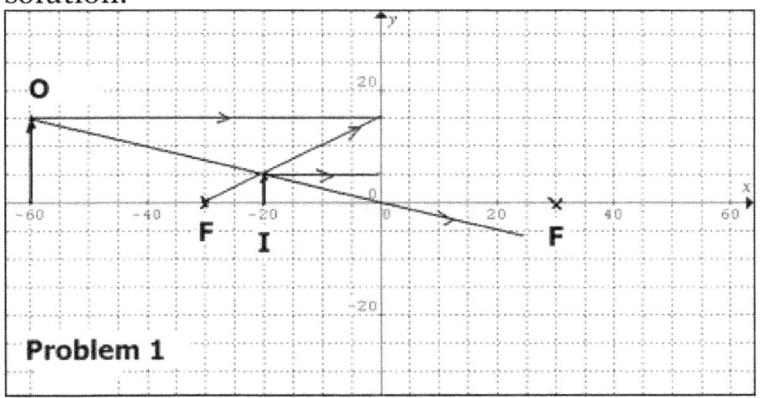

Problem 1

Other Problems:

1. A diverging lens has a focal length of 20 cm.

(a) Where is the image if the object is 20 cm from the lens?

(b) If the object is 4 cm high, how high is the image? (Recall the lateral magnification of a thin lens - with sign conventions in place is:

$M = h'/h = -s'/s$

2. A diverging lens is placed 10 cm from an object and produces an image which is half the size of the object. Find the focal length of the lens.

4. Lens Combinations

FIG. 7: Young Astronomer with home-built refractor for CXC Integrated Science Project

We now look at the practical combination of thin lenses which yields optical instruments such as the one shown in the accompanying photo of Caribbean science student Carson King, who won top price at a Science Exhibition with his self-designed and constructed 2" aperture refracting telescope (which he converted to an astro-camera to take photos of the Moon.)

The typical thin lens combo is shown in the illustrated example, asking *'Where is the final image?'*

The key step is to use the known focal lengths (f1 = 10 cm and f2 = 20 cm) and then perform the working as shown.

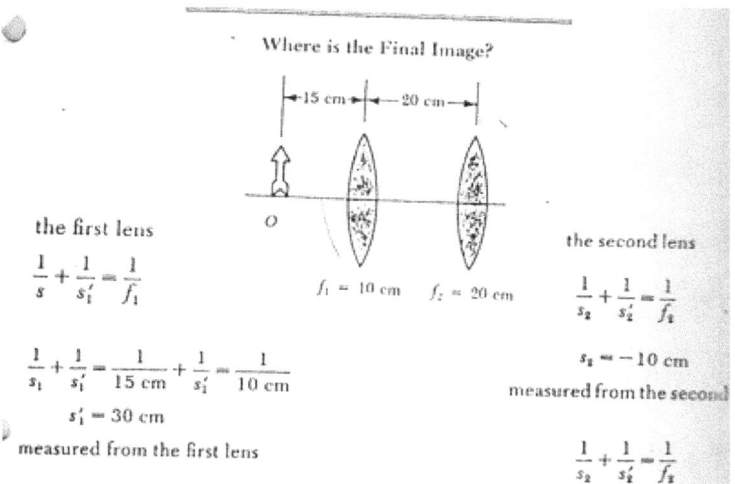

$$\frac{1}{s} + \frac{1}{s'_1} = \frac{1}{f_1}$$

$$\frac{1}{s_1} + \frac{1}{s'_1} = \frac{1}{15 \text{ cm}} + \frac{1}{s'_1} = \frac{1}{10 \text{ cm}}$$

$$s'_1 = 30 \text{ cm}$$

measured from the first lens

$$\frac{1}{s_2} + \frac{1}{s'_2} = \frac{1}{f_2}$$

$s_2 = -10$ cm, measured from the second

$$\frac{1}{s_2} + \frac{1}{s'_2} = \frac{1}{f_2}$$

The trick is to find the image distance s1' for the first lens, then having done that find the object distance s2 of the second lens. One can also obtain the total magnification (lateral) using the magnification formula.

We look at the procedure, then complete the solution for the converging lens system shown.

Procedure for analyzing a thin lens combination:

1) The image of the first lens (L1) is calculated as if the 2nd lens (L2) is not present.

2) The image of the first lens is treated as the object of the 2nd lens.

3) If the image of the first lens lies to the right of the 2nd lens, the image is treated as a virtual object for the 2nd lens (that is, s is negative). Refer again to the sign rules in the previous blog on lenses.

4) The image of the 2nd lens is the final image of the system.

Application :

Using the thin lens eqn. for lens L1:

$1/s_1 + 1/s_1' = 1/15 + 1/s_1' = 1/10$ cm

therefore: **s_1' = 30 cm**

e.g.: $1/s_1' = 1/15 - 1/10 = 5/150$ or $s_1' = 150$ cm/ 5

And for the 2nd lens:

$1/s_2 + 1/s_2' = 1/f_2$

$\rightarrow 1/(-10 \text{ cm}) + 1/s_2' = 1/20$ cm

or $1/s_2' = 1/20$ cm $+ 1/10$ cm $= 30/200$ cm^{-1} or $s_2' = 200/30 = (20/3)$ cm

Thus, the final image lies $(20/3)$ cm **to the right of the 2nd lens**.

The lateral magnification for each lens is defined as before (see, e.g. solutions to previous problems):

$M_1 = (-s_1'/s_1) = -(30 \text{ cm}/15 \text{ cm}) = -2$

$M_2 = (-s_2'/s_2) = -(20/3)\text{cm}/ -10 \text{ cm} = 2/3$

Then the total magnification of the lens system is:

$M_1 M_2 = (-2)(2/3) = -4/3$

So it is:

real, inverted and enlarged by 4/3 times over the object.

In the case of the refracting telescope, such as shown in the photo, the magnifying power is defined by:

$m = F/f(e)$

where F is the focal length of the objective (the main or front lens) and f(e) is the focal length of the eyepiece. Hence, to get a large magnification one needs F to be fairly large and f(e) to be small.

Problems:

(1) A telephoto lens consists of a converging lens of focal length 6 cm placed 4 cm in front of a diverging lens of focal length (-2.5 cm).

a) Do a graphical construction of the system showing where the image would be.
b) Compare the size of the image formed by this combination with the size of the image that would be formed by the positive lens alone.

2) The objective lens of an astronomical telescope has a focal length of 6 ft. The eyepiece has a focal length of

2 inches.

a) Find the angular magnification that the telescope will produce when used for distant objects.

b) A rule for observing extended astronomical objects, such as planets, Moon or nebulae, is that the telescope magnification should not exceed 60 power per inch of objective aperture. The astronomical telescope for this problem is to be used to observe the planet Jupiter. Is the condition met or not? If not, what focal length eyepiece is needed to get the *maximum* angular magnification?

3) The objective of a telescope has a focal length of F = 30 in. When it is used for an object at a great distance, then the distance between the objective and eyepiece is 32 in. What is the angular magnification?

4) On a particular evening the planet Jupiter is reported to be some 40" in angular diameter, (40 arc seconds). A physics student from Harrison College uses a refracting telescope with a 60mm aperture and focal length 700mm to view it. If he uses an eyepiece with a focal length of 4mm, estimate how much greater Jupiter's angular diameter will appear in his telescope.

Spherical Mirrors:

We now examine and investigate image formation using spherical mirrors, either convex or concave. As in the case of convex and concave lenses, light acted upon by any spherical mirror will conform to the laws of geometrical optics, thus we note the following in summary: i) any light ray which passes through the center of curvature (C) of a spherical mirror en route to the mirror surface, will be reflected back upon itself; ii) all light rays which approach the mirror in paths parallel to the optical axis are reflected through a common point on the optical axis known as the principal focus; iii) any light ray which passes through the focal point on its way to the mirror will be reflected parallel to the principal axis.

Investigation techniques for both convex and concave mirrors make use of an optical bench, onto which the mirror can be mounted as well as an 'object' (object-forming illumination device) such as shown below:

Fig.8: Experimental setup for mirror experiments

The object distance, s and the image distance, s', are then varied and the results tabulated, e.g. for differing arrangements of the object, such as: object outside C, object at C, object at f (focal length of lens) or object inside f. For each such case the nature of the image formed is then noted, described, and any other particulars.

In doing these experimental trials the basic law for image formation is: $1/s + 1/s' = 1/f$.

Since the radius of curvature R = 2f, this can also be written as:

$1/s + 1/s' = 2/R$

The summary diagrams applicable to both concave and convex mirrors and their actions are given below:

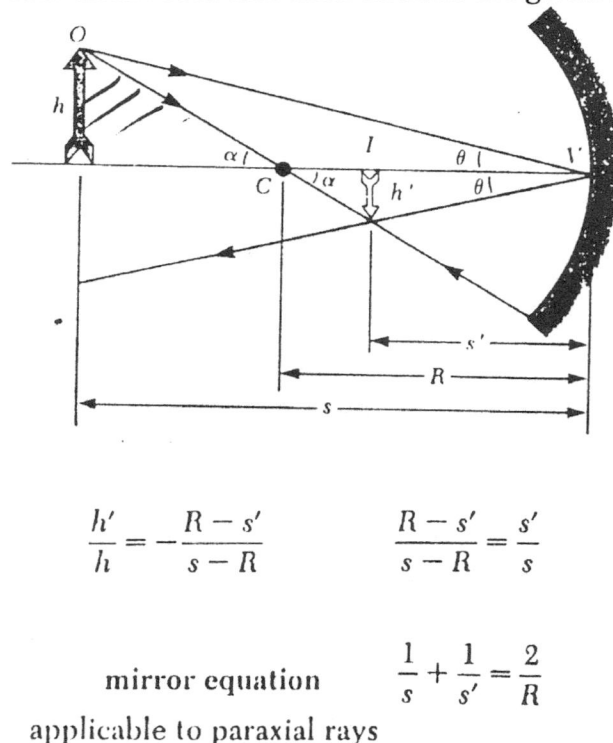

$$\frac{h'}{h} = -\frac{R-s'}{s-R} \qquad \frac{R-s'}{s-R} = \frac{s'}{s}$$

mirror equation $\qquad \dfrac{1}{s} + \dfrac{1}{s'} = \dfrac{2}{R}$

applicable to paraxial rays

Fig. 9: Object within distance s = 2R for concave mirror.

Example Problem:

An experimental setup is used such as shown in Fig. 8 in which the illuminated object is placed 30 cm from a concave mirror. The image is found to be formed at a distance of 120 cm away from the mirror and to have a magnification of 4x. (E.g. h' = 4h). Using this data, find: a) the focal length of the mirror and b) the radius of curvature. Also, (c), sketch a diagram showing the image formation based on the given data.

Solution:

We can use the mirror equation such that:

1/s + 1/s' = 2 /R

Where s = 30 cm and s' = 120 cm, then:

1/30 + 1/ 120 = 2/R or:

R/2 = (120 cm) (30cm)/ [120 cm + 30 cm]

R/ 2 = 3600 cm²/ 150 cm = 24 cm

Since R = 2f then **R/2 = *f*** = 24 cm

So: R = 2 (24 cm) = 48 cm

A sketch of the arrangement is shown below:

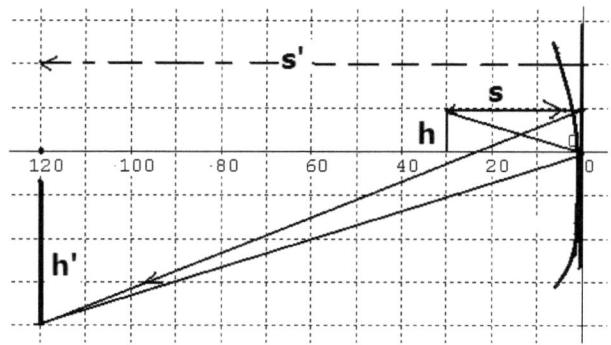

Note in particular that: $4h/h' = 4 = s'/s$.

If an object is very far from the concave mirror, say effectively at infinity ($s = \infty$) then we have a situation peculiar to that for reflecting astronomical telescopes:

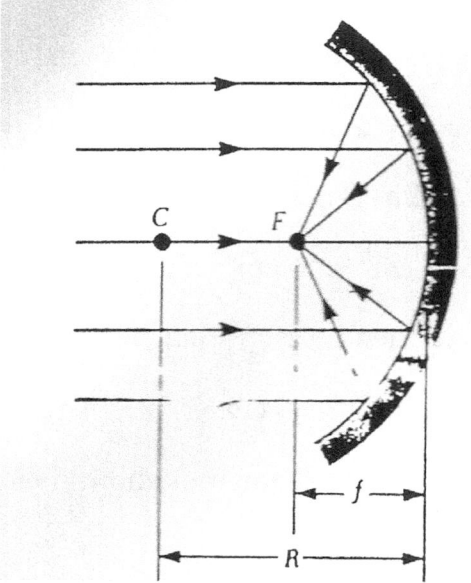

Fig. 10: Object effectively at infinity for concave mirror.

In the case depicted in Fig. 3:

1/s = 1/∞ = 0 and s' = R/2

The mirror equation: 1/s + 1/s' = 1/f

still applies.

Finally, we consider the convex or diverging mirror for which light is reflected from an outer convex surface and the image produced is always erect, virtual and smaller than the real object. This situation is shown below:

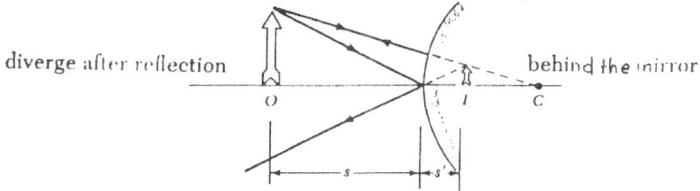

Fig. 11: Image formation for a convex mirror.

The sign conventions for all the possible object, image combinations are summarized below:

Sign Convention for Mirrors
s is + if the object is in front of the mirror (real object).
s is − if the object is in back of the mirror (virtual object).
s' is + if the image is in front of the mirror (real image).
s' is − if the image is in back of the mirror (virtual image).
Both f and R are + if the center of curvature is in front of the mirror (concave mirror).
Both f and R are − if the center of curvature is in back of the mirror (convex mirror).
If M is positive, the image is erect.
If M is negative, the image is inverted.

Problem: The Harry Bayley Observatory's main telescope is shown below:

If the objective focal length is 1000mm, what is the *maximum magnification* it can achieve?

Resolving Power of an Optical Instrument:

Included among the more important applications of diffraction is obtaining the resolution of two sources, say stars. The diagram below (Fig. 13) illustrates the problem for two sources and a single rectangular aperture.

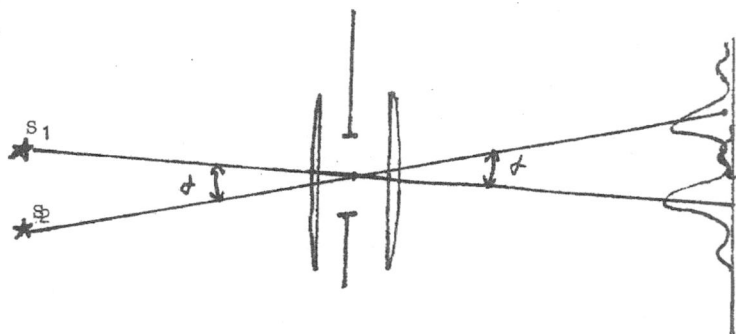

Fig. 13 The resolution of two sources S1 and S2 for a single rectangular aperture.

The resolving power or resolution of an optical instrument just means its ability to distinguish the images of two light sources very near to one another.

In terms of application, it is the diffraction pattern of the sources via the aperture that sets the upper theoretical limit to the resolving power.

In Fig. 13 we have two lenses on either side of a rectangular aperture (slit). Two light sources are detected which are very close together in the line of sight. These produce two diffraction patterns on the screen depicted as two overlapping intensity patterns. Note that the central maxima of the two patterns are separated by an angle α which is the same as the angle (θ) subtended by the two sources at the slit center. Thus, for each pattern, the principal maximum just falls on the second minimum.

By referring back to the intensity distribution sketch at the top of page 209, we see that the angular separation (from center to 2nd minimum) in either pattern would be equal to 2π. Since for a rectangular slit the resolving power may be expressed as:
$\theta_1 = n\lambda / d$

Where n is the number of minima from the center, λ is the wavelength of the light used and d is the slit width.

And in this case we find: $\theta_1 = 2\lambda / d$

For visible light, $\lambda = 5.5 \times 10^{-7}$ m. In performing the calculation, care must be taken to ensure that λ and d are expressed in the same units. For the case depicted in Fig. 10 the two sources are clearly resolved with the angular separation:

$\alpha = 2\theta_1 = 2\lambda / d$

It shouldn't be difficult to see that the degree of resolution deteriorates as one reduces the angular separation. Of particular interest is the case where $\alpha = \pi$, corresponding to the condition $\alpha = \theta_1$, for which the images are just barely resolved. This is known as the Rayleigh criterion and occurs when the central maximum of one diffraction pattern just falls on the first minimum of the other, or:

$\theta_1 = \lambda / d$

Having considered resolution in the context of the single slit Fraunhofer diffraction pattern we're now in position to discuss resolving power as it applies to circular apertures. Such a pattern is arrived at by rotating the single slit about its rotation axis leading to the result shown in Fig. 14:

Fig. 14 Diffraction pattern for circular aperture

Note we again have secondary maxima and minima but these are now observed as concentric rings around a bright central disc. The latter is called **Airy's disc** after Sir George Airy, and is what the observer actually sees when he focuses his telescope on a distant star.

The expression for resolving power is analogous to that for the single rectangular aperture already considered. The chief difference is that n is no longer a whole number. In particular for **the Rayleigh criterion**:

$\theta_1 = 1.22 \ \lambda/D$

where D is now the diameter of the circular aperture. One can see from this that the separation of two light sources will depend very directly on the objective diameter D. The larger D the smaller will be θ_1 and hence the better the resolution.

To fix ideas consider Fig. 15 which shows two stars and their corresponding diffraction patterns for the Rayleigh criterion. One of the stars is on the principal axis, the other is off it. The diagram shows the intensity distributions and (below) what's observed through a telescope.

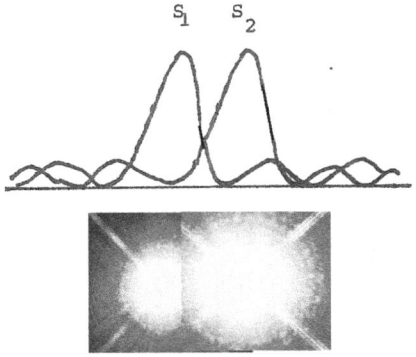

Fig. 15: Diffraction patterns for two stars S1 and S2 at the Rayleigh Criterion.

Practical exercise:

As an exercise we compute the resolving power of the Harry Bayley Observatory Celestron 14 telescope in St. Michael, Barbados, shown below:

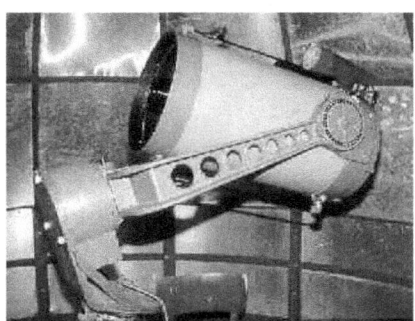

Fig. 14: The Celestron -14 Telescope at the Harry Bayley Observatory, Barbados

Then: $\theta_1 = 1.22 \; \lambda / D$

$\theta_1 = 1.22 \; \lambda / D = 1.22 \; (= 5.5 \times 10^{-7} \text{ m}) / 0.35 \text{ m}$

$\theta_1 = 1.91 \times 10^{-6}$ rad = 0.40 arc sec

Thus, the C-14 telescope will be able to resolve double stars separated by as little as 0.40 seconds of arc.

While the apparent angular diameter of a celestial objects (a discrete object, like a planet) does increase with higher magnification, one does not increase the resolving power at the same time. In effect, one cannot extract any more detail than the diffraction pattern already allows for a star. Further, if one recklessly increases the telescope magnification without regard to the aperture, then one only succeeds in producing a blurred image.

In general, as one increases D, the telescope aperture, the diameter of the Airy disc produced by a distant star is reduced in scale. Thus, we observe star images as smaller and smaller points of light the larger the telescope. What we are really seeing is an enhancement in resolution

Other Problems:

1) The pupil of the eye is approximately 2mm in diameter under moderate illumination. Compute the eye's limit of angular resolution for λ = 550 nm. Is the resolving power greater at night or in broad daylight? Explain.

2) Taking 384,000 km as the mean distance from the Earth to the Moon, find the smallest distance for the separation of two objects on the Moon's surface that can just be resolved by the Harry Bayley Observatory C-14 telescope?

3) A refracting telescope has a 60mm objective with a focal length of 700 mm.

> a) If an eyepiece with a focal length of ¼" is used what magnification can be achieved?
> b) Would this telescope be able to resolve the Sirius A and B double star system if the angular separation is 7."5? Explain.

APPENDICES

APPENDIX I: PHYSICAL CONSTANTS (SI)

Boltzmann constant: $k = 1.3807 \times 10^{-23}$ JK^{-1}

Elementary electronic charge: $e = 1.6022 \times 10^{-19}$ C

Electron mass: $m_e = 9.1094 \times 10^{-31}$ kg

Proton mass: $m_p = 1.6726 \times 10^{-27}$ kg

Gravitational constant: $G = 6.6726 \times 10^{-11}$ m^{3}s^{-2}kg^{-1}

Planck constant: $h = 6.6261 \times 10^{-34}$ J s

$\hbar = h/2\pi = 1.0546 \times 10^{-34}$ J s

Speed of light in vacuum $c = 2.9979 \times 10^8$ ms^{-1}

Permittivity of free space: $\varepsilon_o = 8.8542 \times 10^{-12}$ Fm^{-1}

Permeability of free space: $\mu_0 = 4\pi \times 10^{-7}$ Hm^{-1}

Proton/electron mass ratio: $m_p/m_e = 1.8362 \times 10^3$

Electron charge/mass ratio: $e/m_e = 1.7588 \times 10^{11}$ Ckg^{-1}

Rydberg constant $R_\infty = 1.0974 \times 10^7$ m^{-1}

Bohr radius: $a_0 = 2\ 5.2918 \times 10^{-11}$ m

Fine-structure constant $\alpha = 7.2974 \times 10^{-3}$

Stefan-Boltzmann constant: $\sigma = 5.6705 \times 10^{-8}$ Wm^{-2}K^{-4}

Wavelength associated with 1 eV: $\lambda_0 = hc/e = 1.2398 \times 10^{-6}$ m

Frequency associated with 1 eV: $\nu_0 = 2.4180 \times 10^{14}$ Hz

Energy associated with 1 eV: $h\nu_0 = 1.6022 \times 10^{-19}$ J

Avogadro number $N_A = 6.0221 \times 10^{23}$ mol^{-1}

Gas constant: $R = 8.3145$ JK^{-1}mol^{-1}

Loschmidt's number $n_0 = 2.6868 \times 10^{25}$ m^{-3} (number density at STP)

Atomic mass unit: $u = 1.6605 \times 10^{-27}$ kg

Standard temperature $T_0 = 273.15$ K

Atmospheric pressure $p_0 = 1.0133 \times 10^5$ Pa

Pressure of 1 mm Hg = 1.3332×10^2 Pa (1 torr)

1 calorie (cal) = 4.1868 J

Gravitational acceleration: $g = 9.8067$ ms^{-2}

APPENDIX II: Mathematical Tools

1. Basic Algebra:

$x^a \cdot x^b$ simplifies to $x^{(a+b)}$

$(x \cdot y)^a$ simplifies to $x^a \cdot y^a$

$\left(x^a\right)^b$ simplifies to $x^{(a \cdot b)}$

$\dfrac{1}{x^a}$ simplifies to $x^{(-a)}$

$\dfrac{x^a}{x^b}$ simplifies to $x^{(a-b)}$

$(a+b)^2$ expands to $a^2 + 2 \cdot a \cdot b + b^2$

$a^3 - b^3$
by factoring, yields $(a-b) \cdot \left(a^2 + a \cdot b + b^2\right)$

Solving $a \cdot x^2 + b \cdot x + c = 0$ for x

has solution(s) $\left[\begin{array}{c} \dfrac{1}{(2 \cdot a)} \cdot \left(-b + \sqrt{b^2 - 4 \cdot a \cdot c}\right) \\ \dfrac{1}{(2 \cdot a)} \cdot \left(-b - \sqrt{b^2 - 4 \cdot a \cdot c}\right) \end{array} \right]$

340

1.1 Properties of Logarithms

$\ln(x \cdot y)$ simplifies to $\ln(x) + \ln(y)$

$\ln\left(\dfrac{x}{y}\right)$ simplifies to $\ln(x) - \ln(y)$

$\ln(x^a)$ simplifies to $a \cdot \ln(x)$

2. *Basic Geometry*:

2.1. Parallelogram:

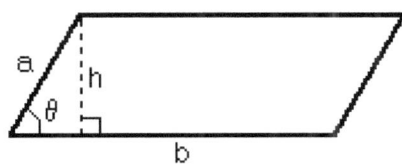

Expressions for Area:

$b \cdot h$ area in terms of b and h

$a \cdot b \cdot \sin(\theta)$ area in terms of a, b, and θ

Perimeter: $2 \cdot a + 2 \cdot b$

2.2 Trapezoid:

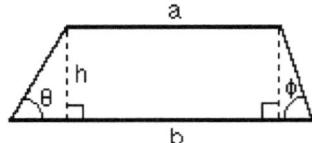

Area: $\frac{1}{2} \cdot h \cdot (a + b)$

Perimeter: $a + b + h \cdot \left(\frac{1}{\sin(\theta)} + \frac{1}{\sin(\phi)} \right)$

2.3 Scalene Triangle:

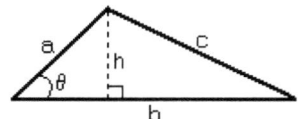

Expressions for Area:

$\frac{1}{2} \cdot h \cdot b$ area in terms of h and b

$\sqrt{s \cdot (s-a) \cdot (s-b) \cdot (s-c)}$ Heron's formula, where s is the semiperimeter

$\frac{1}{2} \cdot a \cdot b \cdot \sin(\theta)$ area in terms of a, b, and θ

Perimeter: $a + b + c$

2.4 Circle & Sector

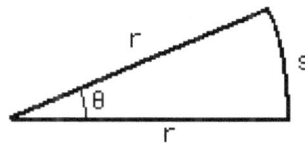

Area: $\pi \cdot r^2$

Area: $\frac{1}{2} \cdot r^2 \cdot \theta$

Perimeter: $2 \cdot \pi \cdot r$

Arc length: $r \cdot \theta$

3.1

Right triangle relationships

$\sin(\theta) = \dfrac{\text{opp}}{\text{hyp}}$

$\cos(\theta) = \dfrac{\text{adj}}{\text{hyp}}$

$\tan(\theta) = \dfrac{\text{opp}}{\text{adj}}$

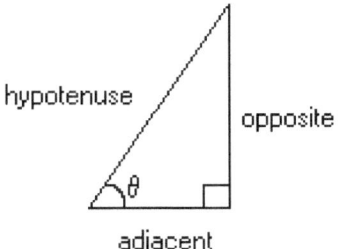

3.1.1 Basic Trigonometric Equalities

$\tan(x)$ simplifies to $\dfrac{\sin(x)}{\cos(x)}$

$\cot(x)$ simplifies to $\dfrac{\cos(x)}{\sin(x)}$

$\sec(x)$ simplifies to $\dfrac{1}{\cos(x)}$

$\csc(x)$ simplifies to $\dfrac{1}{\sin(x)}$

$\sin(\theta)^2 + \cos(\theta)^2$ simplifies to 1

$\tan(x)^2 + 1$ simplifies to $\dfrac{1}{\cos(x)^2}$

$\sin(2 \cdot x)$ expands to $2 \cdot \sin(x) \cdot \cos(x)$

$\cos(x + y)$
expands to $\cos(x) \cdot \cos(y) - \sin(x) \cdot \sin(y)$

$\dfrac{1}{2} \cdot (1 - \cos(2 \cdot t))$ expands to $1 - \cos(t)^2$

$\sin(x \pm y) = \sin x \cos y \pm \cos x \sin y$

$\cos(x \pm y) = \cos x \cos y \pm \sin x \sin y$

3.1.2. Graphs of Trigonometric functions:

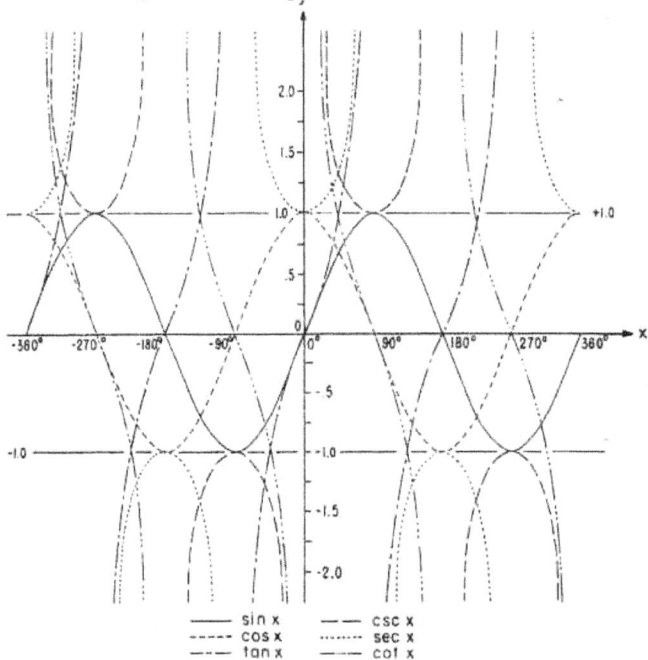

— sin x   — — csc x
- - - - cos x   .... sec x
— · — tan x   — ·· — cot x

3.1.3. Trigonometric functions by quadrant:

	$-\theta$	$\frac{\pi}{2} \pm \theta$	$\pi \pm \theta$	$\frac{3\pi}{2} \pm \theta$	$2k\pi \pm \theta$
sin	$-\sin \theta$	$\cos \theta$	$\mp \sin \theta$	$-\cos \theta$	$\pm \sin \theta$
cos	$\cos \theta$	$\mp \sin \theta$	$-\cos \theta$	$\pm \sin \theta$	$+\cos \theta$
tan	$-\tan \theta$	$\mp \cot \theta$	$\pm \tan \theta$	$\mp \cot \theta$	$\pm \tan \theta$
csc	$-\csc \theta$	$+\sec \theta$	$\mp \csc \theta$	$-\sec \theta$	$\pm \csc \theta$
sec	$\sec \theta$	$\mp \csc \theta$	$-\sec \theta$	$\pm \csc \theta$	$+\sec \theta$
cot	$-\cot \theta$	$\mp \tan \theta$	$\pm \cot \theta$	$\mp \tan \theta$	$\pm \cot \theta$

Where: $0 \leq \theta \leq \pi/2$

3.2.1 Trigonometric Values for Defined Angles:

	0 $0°$	$\pi/12$ $15°$	$\pi/6$ $30°$	$\pi/4$ $45°$	$\pi/3$ $60°$
sin	0	$\frac{\sqrt{2}}{4}(\sqrt{3}-1)$	$1/2$	$\sqrt{2}/2$	$\sqrt{3}/2$
cos	1	$\frac{\sqrt{2}}{4}(\sqrt{3}+1)$	$\sqrt{3}/2$	$\sqrt{2}/2$	$1/2$
tan	0	$2-\sqrt{3}$	$\sqrt{3}/3$	1	$\sqrt{3}$
csc	∞	$\sqrt{2}(\sqrt{3}+1)$	2	$\sqrt{2}$	$2\sqrt{3}/3$
sec	1	$\sqrt{2}(\sqrt{3}-1)$	$2\sqrt{3}/3$	$\sqrt{2}$	2
cot	∞	$2+\sqrt{3}$	$\sqrt{3}$	1	$\sqrt{3}/3$

	$5\pi/12$ $75°$	$\pi/2$ $90°$	$7\pi/12$ $105°$	$2\pi/3$ $120°$
sin	$\frac{\sqrt{2}}{4}(\sqrt{3}+1)$	1	$\frac{\sqrt{2}}{4}(\sqrt{3}+1)$	$\sqrt{3}/2$
cos	$\frac{\sqrt{2}}{4}(\sqrt{3}-1)$	0	$\frac{-\sqrt{2}}{4}(\sqrt{3}-1)$	$-1/2$
tan	$2+\sqrt{3}$	∞	$-(2+\sqrt{3})$	$-\sqrt{3}$
csc	$\sqrt{2}(\sqrt{3}-1)$	1	$\sqrt{2}(\sqrt{3}-1)$	$2\sqrt{3}/3$
sec	$\sqrt{2}(\sqrt{3}+1)$	∞	$-\sqrt{2}(\sqrt{3}+1)$	-2
cot	$2-\sqrt{3}$	0	$-(2-\sqrt{3})$	$-\sqrt{3}/3$

	$3\pi/4$ $135°$	$5\pi/6$ $150°$	$11\pi/12$ $165°$	π $180°$
sin	$\sqrt{2}/2$	$1/2$	$\frac{\sqrt{2}}{4}(\sqrt{3}-1)$	0
cos	$-\sqrt{2}/2$	$-\sqrt{3}/2$	$\frac{-\sqrt{2}}{4}(\sqrt{3}+1)$	-1
tan	-1	$-\sqrt{3}/3$	$-(2-\sqrt{3})$	0
csc	$\sqrt{2}$	2	$\sqrt{2}(\sqrt{3}+1)$	∞
sec	$-\sqrt{2}$	$-2\sqrt{3}/3$	$-\sqrt{2}(\sqrt{3}-1)$	-1
cot	-1	$-\sqrt{3}$	$-(2+\sqrt{3})$	∞

4.1 Basic Differentiation formulae:

$\ln(\,|x|\,)$ by differentiation, yields $\dfrac{1}{x}$

$\sin(x)$ by differentiation, yields $\cos(x)$

$\operatorname{atan}(x)$ by differentiation, yields $\dfrac{1}{\left(1+x^2\right)}$

$\cosh(x)$ by differentiation, yields $\sinh(x)$

4.2 Additional Differentiation formulae:

$$\frac{d}{dz} e^z = e^z$$

$$\frac{d^n}{dz^n} e^{az} = a^n e^{az}$$

$$\frac{d}{dz} a^z = a^z \ln a$$

$$\frac{d}{dz} z^a = a z^{a-1}$$

$$\frac{d}{dz} z^z = (1 + \ln z) z^z$$

4.3 Basic Integration formulae:

$\cos(u)$ by integration, yields $\sin(u)$

$\sec(u) \cdot \tan(u)$ by integration, yields $\dfrac{1}{\cos(u)}$

b^u by integration, yields $\dfrac{1}{\ln(b)} \cdot b^u$

$\dfrac{\sqrt{a^2 + u^2}}{u^2}$ by integration, yields

$$\frac{\left(-\sqrt{a^2 + u^2} + \ln\left(u + \sqrt{a^2 + u^2} \right) \cdot u \right)}{u}$$

4.4 Additional Integration formulae:

$$\int \frac{dz}{z} = \ln z$$

$$\int \ln z \, dz = z \ln z - z$$

$$\int e^{az} dz = e^{az}/a$$

$$\int \sin z \, dz = -\cos z$$

$$\int \cos z \, dz = \sin z$$

$$\int \tan z \, dz = -\ln \cos z = \ln \sec z$$

A PRIMER ON DIFFERENTIAL EQUATIONS

A critical part of pure physics, including astrophysics, is being able to use basic differential equations. This will be especially important when we see how to do essential quantum mechanics as in Appendix III.

Let's start with just about the simplest differential equation imaginable:

dy = x dx

As with all DEs, the solution is accomplished via the process called *integration*. If we integrate both sides (see the previous section), we will obtain:

∫ dy = x ∫ dx, or: $y = x^2/2 + c$

where c is some undefined (as yet) constant of integration. We call the above the "general solution" to the differential equation. This general solution is, in fact, a family of parabolas such as shown in the diagram below.

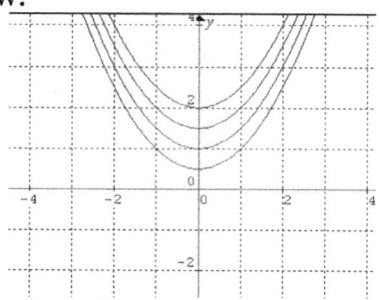

A "family" of parabolas for the equation $y = x^2/2 + c$

A table of values can also be calculated here to expose the limits for x, y entries. (Readers can do this starting with x = 0 and for corresponding dy/dx (or y'). Thus, for x = 0, y' = 0, and for x = +1/2 (or -1/2) then y' = +1/2 (or -1/2) and for x = +1 (or -1) then y' = +1 (or -1) and so forth.)

What the student will find is a graph family such as shown. If we wanted to obtain *the particular solution*, we'd have to have *boundary conditions* available. Usually these designate what values x, y are to have at a particular point, and also often the first derivative (y' or dy/dx) at the same point.

Thus, do we enter the world of first order differential equations of the first degree.

Let's consider the differential equation:

x dx + y dy = 0

One could be understandably tempted to write this in terms of the 1st derivative to obtain:

dy/dx = -x/y

but this serves no useful purpose. It's more productive to directly integrate the equation: xdx = -ydy, viz.

∫x dx = ∫-y dy

(since variables are already separated) to obtain:

x/2 = - y² / 2 , so

$x^2 + y^2 = c$

where c is the constant of integration.

Now, the condition is that y = 2 when x = 1 so:

$(1)^2 + (2)^2 = 1 + 4 = 5$

so the particular solution is: **$x^2 + y^2 = 5$**

Now let's examine a more complex 1st order equation more worthy of the brainpower of the typical readers of the more advanced sections of this book:

Find the general solution and the particular curve passing through the point (0,0) of the differential equation:

$\exp(x) \cos(y) + (1 + \exp(x)) \sin(y) \, dy = 0$

This looks a bit fearsome, but again, the first rule is *simplify*, which means *separating variables* (this is also usually where one's acumen with basic algebra comes in!)

we obtain:

$\exp(x)/(1 + \exp(x)) + [\sin(y)/\cos(y)] \, dy$

$= \exp(x)/(1 + \exp(x)) + \tan(y) dy = 0$

We then integrate: $\int (e^x)/(1 + e^x) \, dx = \int \tan(y) \, dy$

To obtain:

$\ln(1 + e^x) - \ln \cos(y) = \ln c$

Or: $\ln(1 + e^x) = \ln c + \ln \cos(y)$

where again, c is the constant of integration. We can easily simplify the above (using well known principles of natural logs) to get:

$1 + e^x = c \cos(y)$

Which is **the general solution**.

To get the *particular solution* we need to substitute the ordered pair values for (0,0) into the general solution, whence:

$1 + e^0 = c (\cos(0))$

so that: $1 + 1 = c$

and $c = 2$, then we get: $1 + e^x = 2 \cos(y)$

Sample Problems and Solutions:

1) Show that $y = cx^2 - x$ is a solution of the DE: $xy' = 2y + x$

We first take:

$dy/dx = y' = 2cx - 1$

Then substitute for y and y' into the DE:

$xy' = 2y + x$

So: $x(2cx - 1) = 2(cx^2 - x) + x$ (integrating)

$2cx^2 - x = 2cx^2 - 2x + x$

And:

$2cx^2 - x = 2cx^2 - x$

Hence $y = cx^2 - x$ is a solution.

2) For the differential equation: $dy/dx = -x/4y$, sketch the curve which passes through the point (1,1)

We re-arrange to obtain: $-4y\, dy = x dx$

Integrate $-4 \int y\, dy = \int x dx$ to get: $-2y^2 = x^2/2$

Or: $x^2/2 + 2y^2 = c$

One can put in a set of different values of c to generate the family of curves appropriate to the equation. See, e.g. Fig. 1 below.

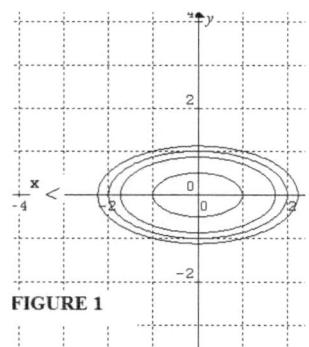

FIGURE 1

On inspection we see *one curve* goes through the point (0, 1) and this is associated with the equation: $x^2/2 + 2y^2 = 2$.

This curve is shown in Fig. 2:

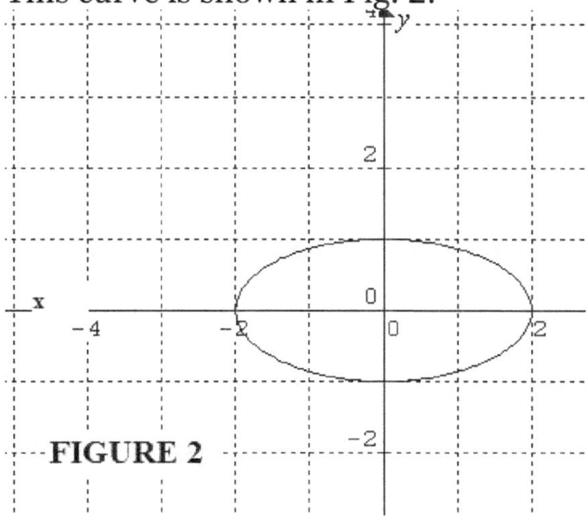

FIGURE 2

3) Find the particular solution for the equation:

exp(x) sec(y)dx + (1 + e^x) sec (y) tan(y) dy = 0

(y = 60 deg when x = 3)

We can simplify and re-arrange the equation to give:

exp(x) dx/ (1 + e^x) = - tan (y) dy

then integrate both sides :

ln(1 + e^x) + ln sec(y) = c

and :

c = (1 + e^x) / cos (y)

Since : $(1 + e^x) = c \cos(y)$

The particular solution is found to be:

$(1 + e^x) = 2(1 + e^3) \cos(y)$

Substitute the values for x, y :

$c = (1 + e^3)/ \cos(60) = (1 + e^3)/ ½$

$c = 2 (1 + e^3)$

Problems:

1. Given that the general solution of:

$dy/dx + y = 2x \exp(-x)$ can be written:

$y = (x^2 + c) \exp(-x)$

solve the initial value problem for y(0) = 2 .

2. Solve the differential equation:

$dy/dx = x^2 \sin(y)$ for initial value, y(-1) = -2

3. Show that $x^3 + 3xy^2 = 1$ is an implicit solution of the differential equation: $(2xy) dy/dx + x^2 + y^2 = 0$

on the interval: $0 \le x \le 1$.

Solutions:

1) If $y = (x^2 + c) \exp(-x)$

And: x = 0 when y = 2, then: 2 = (0 + c) exp(0) = c

Then: y = (x² + 2) exp(-x)

And: y' = -(x² + c)e^(-x) + 2xe^(-x)

2) We first re-arrange to obtain:

dy/ sin(y) = x² dx

which can be integrated (both sides) to yield:

ln(tan (y/2)) = x³/ 3 + c

or ln(tan(y/2)) - x³/3 = c

Now, x = -1 when y = -2 so:

ln (tan(-1)) - (-1)³/3 = c

or:

ln (-π/2) + 1/3 = c

or:

0.443 + π(i) + 1/3 = c

so the solution is: ln(tan (y/2)) = x³/ 3 + 0.443 + π(i) + 1/3

So: ln(tan(y/2)) - x³/3 = 0.776 + πi

3) We see why that interval is given by sketching the curve

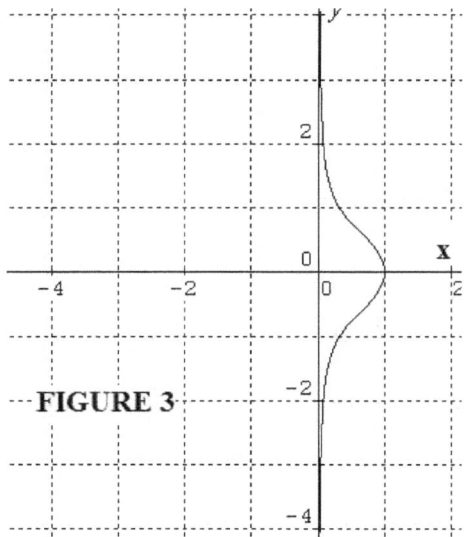

FIGURE 3

It is only well-behaved (continuous) between x slightly larger than 0 and slightly less than 1.

We begin by differentiating implicitly to obtain:

$3x^2 + 3y^2 + 6xy(dy/dx) = 0$

Now divide through by 3:

$x^2 + y^2 + 2xy(dy/dx) = 0$

Then:

$2xy(dy/dx) + x^2 + y^2 = x^2 + y^2 + 2xyy'$

$y' = -(x^2 + y^2)/2xy$

Problem:

Show that $5x^2y^2 - 2x^3y^2 = 1$ is an implicit solution of the differential equation: $x(dy/dx) + y = x^3y^3$

Over the interval $(0, 5/2)$ and sketch the graph.

The graph is sketched in the accompanying diagram (below): the shape shows why the interval must be so rigidly confined.

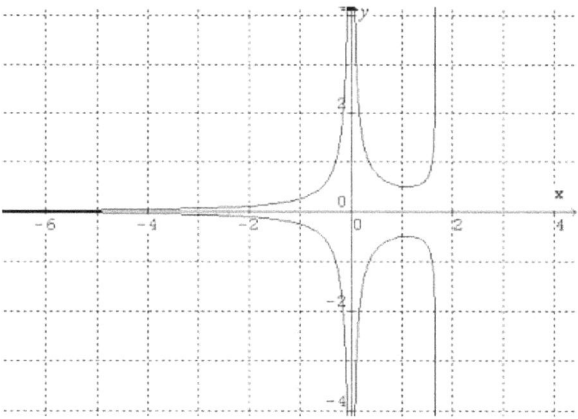

We differentiate implicitly to get:

$10xy^2 + 5x^2 \, 2yy' - 6x^2y^2 + 2x^3 \, 2yy' = 0$

And:

$2yy'(5x^2 + 2x^3) = 6x^2y^2 - 10xy^2$

Then:

$y' = y^2(6x^2 - 10x)/ 2y(5x^2 + 2x^3)$

$= = y(6x^2 - 10x)/ 2(5x^2 + 2x^3)$

But: $= y^2(6x^2 - 10x)/ 2y(5x^2 + 2x^3) = x^2y^3 - y/x$

$= (x^3y^3 - y/x = dy/dx$

So: $x(dy/dx) = x^3y^3 - y$ or $x(dy/dx) + y = = x^3 y^3$

Integrating factors:

This method tarts with writing the typical first order linear DE as:

$dy/dx + Py = Q$

And the name of the game is to account for P and Q and also find **the integrating factor**, r.

Thus, one method for solving the DE shown is to find some function, usually $r = r(x)$ such that if the equation is multiplied by r, the left side becomes the *derivative of the product* ry. That is:

$r(dy/dx) + rPy = rQ$

and we then make the effort to impose upon r the condition that:

$r(dy/dx) + r Py = d/dx (ry)$

which is not always easy, but often can be if one is clever enough!

Expanding the right side of the previous eqn. via differentials:

d/dx (ry) = (rdy + y dr) / dx

and adding to the left, gives:

r(dy/dx) + rPy + (-r (dy/dx) − y (dr/dx)) →

dr/dx = rP

Then, if P = P(x) is a known function, we can solve for r, viz.:

dr/r = Pdx and ln r = ∫ Pdx + ln C

So: r = ± C exp(∫ P dx)

And C can be taken as ± C = 1

Then the function: r = exp(∫ Pdx)

Is called *the integrating factor*

Example:

dy/dx + y = exp(x)

P = 1, Q = exp(x)

Then r = exp(∫dx) = exp(x)

So: exp(x)y = ∫exp(2x) + C = exp(2x)/ 2 + c

And y = exp(x)/2 + C exp(-x) or:

$y = e^x/2 + Ce^{-x}$

Problems:

(1) Solve: $x^2y\, dy - xy^2\, dx - x^3y^2 dx = 0$

(2) Solve using any method for integrating factors:

$x\, (dy/dx) - 3y = x^2$

Solutions:

(1) This is easy once one has access to a table of differentials or is able to work them out!

Factor to obtain:

$xy(xdy - ydx) - x^2 y^2\, dx = 0$

Now, multiply by $(x^{-2}y^{-2})$:

$(x\, dy - ydx)/xy - x\, dx = 0$

Then by applying the property of the differential, e.g. $d(\ln y/x)$:

$d(\ln y/x) - xdx = 0$

Integrating: $\ln(y/x) - x^2/2 = c$

(2) Put the equation into the form:

$dy/dx + Py = Q$

Then: $dy/dx - 3y/x = x$

So: $P = (-3/x)$ and $Q = x$

Therefore:

$r = \exp(\int P dx) = \exp(-3 \ln x) = 1/e^{3 \ln x} = 1/x^3$

Therefore:

$(1/x^3) y = \int_x (x/x^3) dx + C = -1/x + C$

So: $y = -x^2 + Cx^3$

Partial derivatives and Exact Differential Equations:

Definition:

Given some differential equation:

$M(x,y)dx + N(x,y)dy = 0$

If there exists a function $f(x,y)$ such that:

$\partial f/\partial x = M(x,y)$ and $\partial f/\partial y = N(x,y)$

Then the differential equation is said to be exact. (where $\partial f/\partial x$ and $\partial f/\partial y$ are the *partial derivatives of f with respect to x and y, respectively*)

A necessary and sufficient condition that the DE is exact is also that:

$\partial M/\partial y = \partial N/\partial x$

As an example, we want to find out if:

$dy/dx = (x + y)/xy^2$

is exact. Re-arrange to get:

$(x + y)dx - xy^2 dy = 0$

Then: $M(x,y) = (x + y)$

$N(x, y) = (-xy^2)$

The partial derivative with respect to x is taken by simply differentiating with respect to x, holding y constant. The partial derivative with respect to y is taken by differentiating with respect to y, holding x constant.

Thus: $\partial M/\partial y = x$ and $\partial N/\partial x = -2yx$

So this DE is **not exact** since: $\partial M/\partial y \neq \partial N/\partial x$

Example (2) Show that the DE:

$2xy\,dx + (1 + x^2)\,dy = 0$

is exact and find the general solution

As before: $M(x,y) = 2xy$ and $N(x.y) = (1 + x^2)$

Then: $\partial M/\partial y = 2x$ and $\partial N/\partial x = 2x$

So yes, the DE *is* exact!

To obtain the general solution, let:

$f(x,y) = \int_x (2xy \, dx + c(y)) = x^2y + c(y)$

since $\partial f/\partial y = N$ we have:

$\partial/\partial y \, [\, x^2y + c(y)] = x^2 + d/dy \, [c(y)] = 1 + x^2$

(Since: $\partial/\partial y \, [\, x^2y + c(y)] = x^2$ and $\partial c(y)/\partial y = 0$)

We see: $d/dy \, [c(y)] = 1$ and $c(y) = y$. e.g.

$\int dc(y) = c(y) = \int dy = y$

The function (or *general solution*) is then:

$f(x,y) = x^2y + c(y) = x^2y + y$ or $x^2y + y = c$

Problems:

(1) State whether the DE: $dy/dx = (2x + y^2)/ -2xy$ is exact

(2) Show that the DE below is exact and find the solution

$(3xy^4 + x)dx + (6x^2y^3 - 2y^2 + 7)dy = 0$

Solutions:

(1) We first re-arrange to obtain:

$(-2xy) \, dy = (2x + y^2) \, dx$

And: $(2x + y^2) \, dx + (2xy) dy = 0$

Then: $M = (2x + y^2)$ and $N = (2xy)$

Take partial derivatives $\partial M/\partial y$ and $\partial N/\partial x$:

$\partial M/\partial y = 2y$

$\partial N/\partial x = 2y$

Since the two partials are equal, the DE is exact.

(2) First, take the partials to make sure it's exact:

$M = (3xy^4 + x)$ and $\partial M/\partial y = 12xy^3$

$N = 6x^2y^3 - 2y^2 + 7)$ and $\partial N/\partial x = 12xy^3$

So, it's exact.

Now, let:

$f(x,y) = \int_x (3xy^4 + x) \, dx + C(y)$

$= 3x^2y^4/2 + x^2/2 + C(y)$

Then:

$dC/dy = 6x^2y^3 - 2y^2 + 7 - 6x^2y^3$

Or: $dC/dy = -2y^2 + 7$

Therefore:

$\int dC(y) = C(y) = \int (-2y^2 + 7) \, dy = -2y^3/3 + 7y$

So the general solution is:

$f(x,y) = 3x^2y^4/2 + x^2/2 - 2y^3/3 + 7y$

Additional Problem:

Consider the differential equation:

$(3x^2 y + 2) \, dx + (x^3 + y) \, dy = 0$

Determine whether it is an exact DE or not. If it is, find the general solution *and then the particular* for an initial condition such that: $y(1) = 3$.

We have: $M = (3x^2 y + 2)$ and $\partial M/\partial y = 3x^2$

And: $N = (x^3 + y)$ and $\partial N/\partial x = 3x^2$

So, the DE is exact.

We have then:

$f(x,y) = \int_x (3x^2 y + 2) \, dx + C(y) =$

$(x^3 y + 2x) + C'(y)$

$\int_y dC(y) dy = C(y) = \int_y (x^3 + y) \, dy + c_1 = y^2/2 + c_1$

This solution satisfies: $f(x,y) = c_2$

Or: $x^3 y + 2x + y^2/2 + c_1 = c_2 = \int_x (3x^2 y + 2) \, dx + C(y)$

So the final general soln. is: : $x^3y + 2x + y^2/2 + c = 0$

To satisfy the condition $y(1) = 3$:

$(1)^3 + 2(1) + (3)^2/2 + c = 0$

So: $c = -9\frac{1}{2}$

Therefore: : $x^3y + 2x + y^2/2 - 19/2 = 0$

Note:

Any differential equation of the form:

$M(x,y)\,dx + N(x,y)\,dy = 0$

   Can be transformed into an exact DE by multiplying it by some suitable factor, call it $r(x,y)$. This makes the DE "exact" and the factor that makes it so is called an *"integrating factor"*. (See previous section on general integrating factors to do with the ordinary differential equations). *Usually* an appropriate $r(x,y)$ can be found on inspection of the DE and visualizing how it might be most directly simplified, say if both sides were multiplied through by some expression.

Problem for readers:

Find an integrating factor for the DE:

$x\,dy + y\,dx = x^2y^2\,dx$ and solve.

The SI units:

TABLE I : BASE UNITS (S.I.)

UNIT NAME	Dimension SYMBOL	S.I. UNIT
Mass	M	kg
Length	L	m
Time	T	s
Current	I	A
Temperature	T	C°
Amount of subs.	n	mol

One can also obtain a set of derived units from the basic:

TABLE II : DERIVED UNITS (S.I.)

Derived Unit	Dimensions	S.I. Symbol
Area	L^2	m^2
Volume	L^3	m^3
Density	$M L^{-3}$	$kg\ m^{-3}$
Velocity	$L T^{-1}$	$m\ s^{-1}$
Acceleration	$L T^{-2}$	$m\ s^{-2}$
Force	$M L T^{-2}$	$kg\ m\ s^{-2}$ (N)
Momentum	$M L T^{-1}$	$kg\ m\ s^{-1}$
Pressure	$M L^{-1} T^{-2}$	$N\ m^{-2}$

APPENDIX III:

Astronomical-Astrophysical Data

The H.R. Diagram

M_V

Sp	Super-giants Ia	Super-giants Ib	Bright giants II	Giants III	Sub-giants IV	Main seq. dwarfs V	ZAMS V	White dwarfs VII	Pop II. Sub-dwarfs VI	Pop II. Red branch	Pop II. Horiz. branch
O5	−6.4			−5.4		−5.7					
B0	−6.7	−6.1	−5.4	−5.0	−4.7	−4.1	−3.3	+10.2			
B5	−6.9	−5.7	−4.3	−2.4	−1.8	−1.1	−0.2	+10.7			
A0	−7.1	−5.3	−3.1	−0.2	+0.1	+0.7	+1.5	+11.3			+2.3
A5	−7.7	−4.9	−2.6	+0.5	+1.4	+2.0	+2.4	+12.2			+0.8
F0	−8.2	−4.7	−2.3	+1.2	+2.0	+2.6	+3.1	+12.9			+0.5
F5	−7.7	−4.7	−2.2	+1.4	+2.3	+3.4	+3.9	+13.6	+4.8	+4.8	+0.4
G0	−7.5	−4.7	−2.1	+1.1	+2.9	+4.4	+4.6	+14.3	+5.7	+4.1	+0.3
G5	−7.5	−4.7	−2.1	+0.7	+3.1	+5.1	+5.2	+14.9	+6.4	+2.0	−0.1
K0	−7.5	−4.6	−2.1	+0.5	+3.2	+5.9	+6.0	+15.3	+7.3	−0.2	−0.6
K5	−7.5	−4.6	−2.2	−0.2		+7.3	+7.3	+15	+8.4	−2.2	−2.2
M0	−7.5	−4.6	−2.3	−0.4		+9.0	+9.0	+15	+10	−3	−3
M2	−7		−2.4	−0.6		+10.0	+10.0		+12		
M5				−0.8		+11.8	+11.8		+14		
M8						+16			+16		

Relation between spectral type (Sp) and Absolute magnitude (M_V) for selected stellar classes-populations on the H-R diagram

THE TEN PRIMARY STANDARD STARS OF THE U, B, V SYSTEM

Name	V	B−V	U−B	Sp. Type
α Ari	2.00	+1.151	+1.12	K2 III
HR 875	5.17	+0.084	+0.05	A1 V
β Cnc	3.52	+1.480	+1.78	K4 III
η Hya	4.30	−0.195	−0.74	B3 V
β Lib	2.61	−0.108	−0.37	B8 V
α Ser	2.65	+1.168	+1.24	K2 III
ε CrB	4.15	+1.230	+1.28	K3 III
τ Her	3.89	−0.152	−0.56	B5 IV
10 Lac	4.88	−0.203	−1.04	O9 V
HR 8832	5.57	+1.010	+0.89	K3 V

Note: A "Primary standard star" is as the name implies – a referenced standard, for the spectral indices V, B- V, U − B.

STELLAR MASS, LUMINOSITY, RADIUS AND DENSITY

Luminosity and radius with mass, white dwarfs omitted

$\log(\mathcal{M}/\mathcal{M}_\odot)$	M_{bol}	$\log(\mathcal{L}/\mathcal{L}_\odot)$	M_V	M_B	$\log(\mathcal{R}/\mathcal{R}_\odot)$ main seq.
−1.0	+12.1	−2.9	15.5	+17.1	−0.9
−0.8	+10.9	−2.5	13.9	+15.5	−0.7
−0.6	+9.7	−2.0	12.2	+13.9	−0.5
−0.4	+8.4	−1.5	10.2	+11.8	−0.3
−0.2	+6.6	−0.8	7.5	+8.7	−0.14
0.0	+4.7	0.0	4.8	+5.5	0.00
+0.2	+2.7	+0.8	2.7	+3.0	+0.10
+0.4	+0.7	+1.6	1.1	+1.1	+0.32
+0.6	−1.1	+2.3	−0.2	−0.1	+0.49
+0.8	−2.9	+3.0	−1.1	−1.2	+0.58
+1.0	−4.6	+3.7	−2.2	−2.4	+0.72
+1.2	−6.3	+4.4	−3.4	−3.6	+0.86
+1.4	−7.6	+4.9	−4.6	−4.9	+1.00
+1.6	−8.9	+5.4	−5.6	−6.0	+1.15
+1.8	−10.2	+6.0	−6.3	−6.9	+1.3

Mass, radius, luminosity, and mean density with spectral class

I = supergiant, III = giant, V = dwarf

A single column between III and V represents main sequence

Sp	$\log(\mathcal{M}/\mathcal{M}_\odot)$			$\log(\mathcal{R}/\mathcal{R}_\odot)$			$\log(\mathcal{L}/\mathcal{L}_\odot)$			$\log\bar{\rho}$		
	I	III	V	I	III	V	I	III	V	I	III	V
O5	+2.2		+1.6			+1.25			+5.7			−2.0
B0	+1.7		+1.25	+1.3	+1.2	+0.87	+5.4		+4.3	−2.1		−1.2
B5	+1.4		+0.81	+1.5	+1.0	+0.58	+4.8		+2.9	−2.9		−0.78
A0	+1.2		+0.51	+1.6	+0.8	+0.40	+4.3		+1.9	−3.5		−0.55
A5	+1.1		+0.32	+1.7		+0.24	+4.0		+1.3	−3.8		−0.26
F0	+1.1		+0.23	+1.8		+0.13	+3.9		+0.8	−4.2		−0.01
F5	+1.0		+0.11	+1.9	+0.6	+0.08	+3.8		+0.4	−4.5		+0.03
G0	+1.0	+0.4	+0.04	+2.0	+0.8	+0.02	+3.8	+1.5	+0.1	−4.9	−1.8	+0.13
G5	+1.1	+0.5	−0.03	+2.1	+1.0	−0.03	+3.8	+1.7	−0.1	−5.2	−2.4	+0.20
K0	+1.1	+0.6	−0.11	+2.3	+1.2	−0.07	+3.9	+1.9	−0.4	−5.7	−2.9	+0.25
K5	+1.2	+0.7	−0.16	+2.6	+1.4	−0.13	+4.2	+2.3	−0.8	−6.4	−3.4	+0.38
M0	+1.2	+0.8	−0.33	+2.7		−0.20	+4.5	+2.6	−1.2	−6.7	−4	+0.4
M2	+1.3		−0.41	+2.9		−0.3	+4.7	+2.8	−1.5	−7.2		+0.7
M5			−0.67			−0.5		+3.0	−2.1			+1.0
M8			−1.0			−0.9			−3.1			+1.8

Note carefully, the attributes that apply to the Main Sequence, because these represent those that can be easily plotted on an H-R diagram.

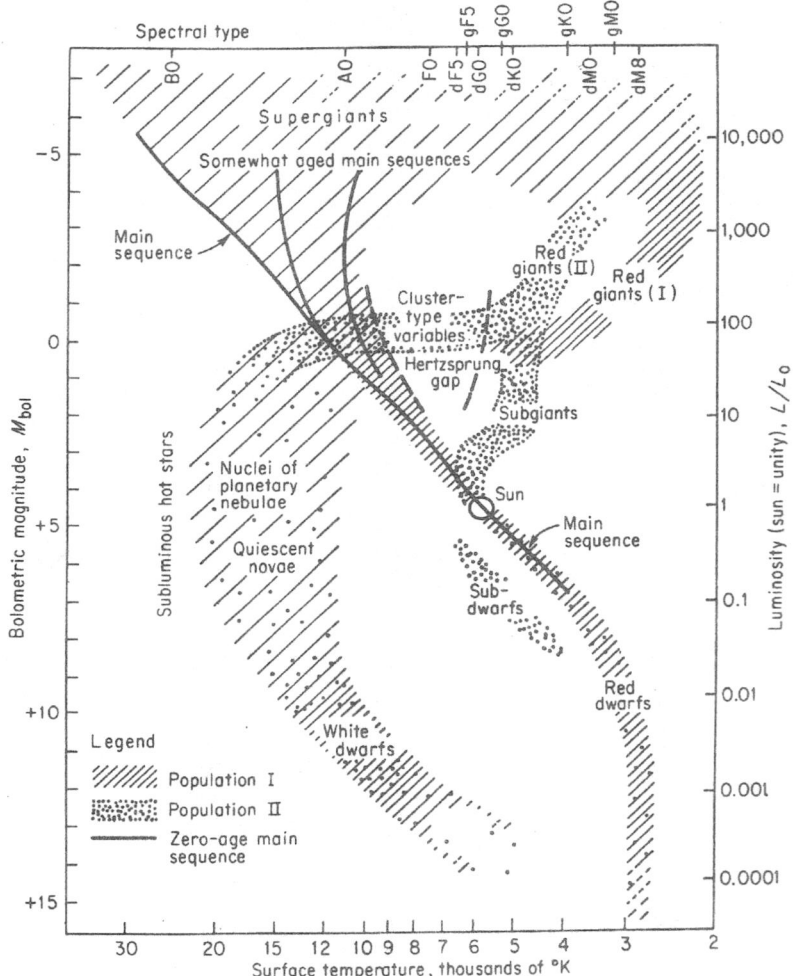

The Hertzsprung-Russell (H-R) Diagram with main stellar groups, populations shown.

Selected Astrophysical Constants

Solar Mass $M = 1.989 \times 10^{30}$ kg

Solar radius $R = 6.9599 \times 10^8$ m

Solar Luminosity $L = 3.826 \times 10^{26}$ Watts (J/s)

Astronomical Unit (AU) $= 1.495979 \times 10^{11}$ m

Hubble Constant $H_o = 100$ km sec^{-1} Mpc^{-1}

Hubble time: $t_H = 1/H_o = 13.7 \times 10^9$ yrs.

Solar effective temperature $T_e = 5800$ K

Apparent (photometric) solar magnitude: -26.73

Thomson cross section:

$\sigma_e = 6.65 \times 10^{-25}$ cm^2

Magnetic diffusivity (Solar): $\eta = 327.6$ m^2/s

Bohr magneton:

$e\hbar/2m_e = 9.27400968(20) \times 10^{-24}$ JT^{-1}

Comprehensive Data: Terrestrial Planets:

	Mercury	Venus	Earth	Mars
Mean radius R_V (km)	2440 ± 1	6051.8(4 ± 1)	6371.0(1 ± 2)	3389.9(2 ± 4)
Mass ($\times 10^{23}$ kg)	3.302	48.685	59.736	6.4185
Volume ($\times 10^{10}$ km^3)	6.085	92.843	108.321	16.318
Density (g cm^{-3})	5.427	5.204	5.515	3.933(5 ± 4)
Flattening f			1/298.257	1/154.409
Semimajor axis			6378.136	3397 ± 4
Siderial rotation period	58.6462d	−243.0185d	23.93419hr	24.622962hr
Rotation rate ω ($\times 10^5$s)	0.124001	−0.029924	7.292115	7.088218
Mean solar day (in days)	175.9421	116.7490	1.002738	1.0274907d
$m_V = \omega^2 R_V^3/GM$	10×10^{-7}	61×10^{-3}	0.0034498	0.0045699
Polar gravity (m s^{-2})			9.832186	3.758
Equatorial gravity (m s^{-2})	3.701	8.870	9.780327	3.690
Moment of inertia: I/MR_o^2	0.33	0.33	0.3308	0.366
Core radius (km)	~1600	~3200	3485	~1700
Potential Love no. k_2		~0.25	0.299	~0.14
Grav. spectral factor: u ($\times 10^5$)		1.5	1.0	14
Topo. spectral factor: t ($\times 10^5$)		23	32	96
Figure offset($R_{CF} - R_{CM}$) (km)		0.19 ± 01	0.80	2.50 ± 0.07
Offset (lat./long.)		11°/102°	46°/35°	62°/88°
Planetary Solar constant (W m^2)	9936.9	2613.9	1367.6	589.0
Mean Temperature (K)		735	270	210
Atmospheric Pressure (bar)		90	1.0	0.0056
Maximum angular diameter	11″0	60″2		17″9″
Visual magnitude V(1,0)	−0.42	−4.40	−3.86	−1.52
Geometric albedo	0.106	0.65	0.367	0.150
Obliquity to orbit (deg)	~0.1	177.3	23.45	25.19
Sidereal orbit period (yr)	0.2408445	0.6151826	0.9999786	1.88071105
Sidereal orbit period (day)	87.968435	224.695434	365.242190	686.92971
Mean daily motion: n (° d^{-1})	4.0923771	1.6021687	0.9856474	0.5240711
Orbit velocity (km s^{-1})	47.8725	35.0214	29.7859	24.1309
Escape velocity v_∞ (km s^{-1})	4.435	10.361	11.186	5.027

Comprehensive Data: Giant Planets:

	Jupiter	Saturn	Uranus	Neptune
Mass (10^{24} kg)	1898.6	568.46	86.832	102.43
Density (g cm^{-3})	1.326	0.6873	1.318	1.638
Equatorial radius (1bar) a (km)	71492 ± 4	60268 ± 4	25559 ± 4	24766 ± 15
Polar radius b (km)	66854 ± 10	54364 ± 10	24973 ± 20	24342 ± 30
Volumetric mean radius: R_V (km)	69911 ± 6	58232 ± 6	25362 ± 12	24624 ± 21
flattening $f = (a-b)/a$	0.06487	0.09796	$^A 0.02293$	0.0171
	± 0.00015	± 0.00018	± 0.0008	± 0.0014
Rotation period: T_{mag}	$9^h 55^m 27\overset{s}{.}3$	$10^h 39^m 22\overset{s}{.}4$	17.24 ± 0.01 h	16.11 ± 0.01 h
Rotation rate ω_{mag} (10^{-4} rad s^{-1})	1.75853	1.63785	1.012	1.083
$m = \omega^2 a^3 / GM$	0.089195	0.15481	0.02954	0.02609
Hydrostatic flattening f_h ^{B}	0.06509	0.09829	0.01987	0.01804
Inferred rotation period T_h (hr)	9.894 ± 0.02	10.61 ± 0.02	17.14 ± 0.9	16.7 ± 1.4
$k_s = 3 J_2/m$	0.494	0.317	0.357	0.407
Moment of inertia: $I/M R_e^2$ ^{C}	0.254	0.210	0.225	
$I/M R^2_o$ (upper bound) ^{D}	0.267	0.231	0.232	0.239
Rocky core mass (M_c/M) ^{C}	0.0261	0.1027	0.0012	
Y factor (He/H ratio)	0.18 ± 0.04	0.06 ± 06	0.262 ± 0.048	0.235 ± 0.040
Equatorial gravity g_e (m s^{-2})	23.12 ± 0.01	8.96 ± 0.01	8.69 ± 0.01	11.00 ± 0.05
Polar gravity g_p (m s^{-2})	27.01 ± 0.01	12.14 ± 0.01	9.19 ± 0.02	11.41 ± 0.03
Geometric albedo	0.52	0.47	0.51	0.41
Visual magnitude $V(1,0)$	-9.40	-8.88	-7.19	-6.87
Visual magnitude (opposition)	-2.70	$+0.67$	$+5.52$	$+7.84$
Obliquity to orbit (deg)	3.12	26.73	97.86	29.56
Sidereal orbit period (yr)	11.856523	29.423519	83.747407	163.72321
Sidereal orbit period (day)	4330.595	10746.940	30588.740	59799.900
Mean daily motion n (° d^{-1})	0.0831294	0.0334979	0.0117690	0.0060200
Mean orbit velocity (km s^{-1})	13.0697	9.6624	5.4778	4.7490
Atmospheric temperature (1 bar) (K)	165 ± 5	134 ± 4	76 ± 2	72 ± 2
Heat flow/Mass ($\times 10^7$ erg g^{-1} s^{-1})	15	15	0.6 ± 0.6	2
Planetary solar constant (W m^{-2})	50.5	15.04	3.71	1.47
Mag. dipole moment (gauss-R_p^3)	4.2	0.21	0.23	0.133
Dipole tilt/offset (deg/R_p)	9.6/0.1	0.0/0.0	58.6/0.3	47/0.55
Escape velocity v (km s^{-1})	59.5	35.5	21.3	23.5

Planetary Mean Orbital Data (1):

Planet	A AU $AU\ Cy^{-1}$	e Cy^{-1}	I deg $''Cy^{-1}$	Ω deg $''Cy^{-1}$	$\tilde{\omega}$ deg $''Cy^{-1}$	L deg $''Cy^{-1}$
Mercury	0.38709893	0.20563069	7.00487	48.33167	77.45645	252.25084
	0.00000066	0.00002527	−23.51	−446.30	573.57	538101628.29
mean	0.38709880	0.20563175	7.00499	48.33089	77.45612	252.25091
orbit		0.00002041	−21.43	−451.52	571.91	538101628.89
Venus	0.72333199	0.00677323	3.39471	76.68069	131.53298	181.97973
	0.00000092	−0.00004938	−2.86	−996.89	−108.80	210664136.06
	0.72333201	0.00677177	3.39447	76.67992	131.56371	181.97980
		−0.00004777	−3.08	−1000.85	17.55	21066136.43

Planetary Mean Orbital Data (2):

Planet	A AU $AU\ Cy^{-1}$	e Cy^{-1}	I deg $''Cy^{-1}$	Ω deg $''Cy^{-1}$	$\tilde{\omega}$ deg $''Cy^{-1}$	L deg $''Cy^{-1}$
Earth	1.00000011	0.01671022	0.00005	−11.26064	102.94719	100.46435
	−0.00000005	−0.00003804	−46.94	−18228.25	1198.28	129597740.63
	1.00000083	0.016708617	0.0	0.0	102.93735	100.46645
		−0.00004204	−46.60	−867.93	1161.12	129597742.28
Mars	1.52366231	0.09341233	1.85061	49.57854	336.04084	355.45332
	−0.00007221	0.00011902	−25.47	−1020.19	1560.78	68905103.78
	1.52368946	0.09340062	1.84973	49.55809	336.60234	355.43327
		0.00009048	−29.33	−1062.90	1598.05	68905077.49
Jupiter	5.20336301	0.04839266	1.30530	100.55615	14.75385	34.40438
	0.00060737	−0.00012880	−4.15	1217.17	839.93	10925078.35
	5.20275842	0.04849485	1.30327	100.46444	14.33131	34.35148
		0.00016322	−7.16	636.20	777.88	10925660.38
Saturn	9.53707032	0.05415060	2.48446	113.71504	92.43194	49.94432
	−0.00301530	−0.00036762	6.11	−1591.05	−1948.89	1401052.95
	9.54282442	0.05550862	2.48888	113.66552	93.05678	50.07747
		−0.00034664	9.18	−924.02	2039.55	4399609.86
Uranus	19.19126393	0.04716771	0.76986	74.22988	170.96424	313.23218
	0.00152025	−0.00019150	6.11	−1591.05	−1948.89	1513052.95
	19.19205970	0.04629590	0.77320	74.00595	173.00516	314.05501
		−0.00002729	−6.07	266.91	321.56	1542481.19
Neptune	30.06896348	0.00858587	1.76917	131.72169	44.97135	304.88003
	−0.00125196	0.00002514	−3.64	−151.25	−844.43	786449.21
	30.06893043	0.00898809	1.76995	131.78406	48.12369	304.34867
		0.00000603	8.12	−22.19	105.07	786550.32
Pluto	39.48168677	0.24880766	17.14175	110.30347	224.06676	238.92881
	−0.00076912	0.00006465	11.07	−37.33	−132.25	522747.90

APPENDIX IV. Nuclear Fusion Reactions:

In general a nuclear fusion reaction is one in which two light nuclei combine (fuse) to form a heavier nucleus with positive energy given off (the Q of the reaction). Nuclear fusion is demonstrated in its most compelling form in the case of stellar energy. Exhaustive investigations in this regard, eventually led to the realization that fusion was the only practical energy by which stars could be sustained over long periods of time, such as billions of years.

In the Sun, for example, two distinct nuclear fusion processes occur: 1) the proton-proton cycle, and 2) the carbon-nitrogen cycle.

In the first of these (the easier one because it has fewer reactions):

$$^1H + {}^1H + e^- \rightarrow {}^2H + \nu + 1.44 \text{ MeV}$$

$$^2D + {}^1H \rightarrow {}^3He + \gamma + 5.49 \text{ MeV}$$

$$^3He + {}^3He \rightarrow {}^4He + {}^1H + {}^1H + 12.85 \text{ MeV}$$

The top line shows two protons fusing to yield deuterium (heavy hydrogen) with a positron and neutrino (ν) emitted, along with 1.44 MeV of energy. Empirical evidence of this reaction is obtained from gallium detectors, of the neutrinos given off, which are within 1-2% of what theoretical models predict.[3] In the second fusion reaction, the deuterium combines with a proton to give the isotope helium 3, along with a gamma ray (γ) and 5.49 MeV energy. In the final fusion, two helium-3 nuclei combine to yield one helium-4 nucleus, along with two protons, and 12. 85 MeV energy. Note that the two ending product protons commence the cycle anew, so that the generation of nuclear energy is ongoing.

The ending quantities on the right sides of each part of the cycle denote the Q of the reaction for that part. Let us check the Q for the first and simplest part. We know the hydrogen mass = **1.007825 u and for deuterium we have (from atomic tables): : ^{2}D = = 2.01410 u. Then:**

Q = [2(1.007825 u) − 2.01410 u] c^2

Q = [2.01565 u − 2.015941u] c^2

Q = [2.01565 − 2.01410] 931.5 MeV/u

Q = [0.00155] 931.5 MeV/u = **1.44 MeV**

The effect of ongoing fusion reactions such as this, means that the central core of the Sun becomes heavier and heavier, as more and more helium is produced. This despite the fact that the Sun as a whole is losing an amount of mass of roughly 4 x 10^6 metric tones per second

---

[3] See, e.g. **Physics Today**: Reports, April, 1995, p. 19.

Insight Problem:

If the atomic mass for helium 3 (^{3}**He**) is equal to 3.01603 u, then verify the other Q-values for the last two parts of the proton-proton cycle. A simplified, compressed "net reaction":

^{1}H + ^{1}H + ^{1}H + ^{1}H → ^{4}He + Energy

This is sometimes used to evaluate the total energy released in the proton-proton cycle. Compute this energy and compare to the value obtained for the total energy released in the earlier example. Can you account for the difference?

Nuclear Fusion Reactions in the Aging Sun:

At some stage, when nearly the entire solar core is helium a new helium fusion phase will be ushered in (at higher temperature), such that the following reaction series, known as the 'triple alpha' process, kicks in:

^{4}He + ^{4}He → ^{8}Be + γ (- 95 keV)

^{8}Be + ^{4}He → ^{12}C + γ + 7.4 MeV

Here, the two alpha particles (helium nuclei) first fuse to give unstable beryllium and a gamma ray (γ), with 95 keV energy *absorbed*. Then the beryllium fuses with a helium-4 to give carbon–12 plus a gamma ray and 7.4 MeV energy given off.

In this way a new cycle commences, leading to a heavier molecular weight core. Each successive burning phase, however, is less efficient than its predecessor, as can be seen by comparing the energy given off in the triple alpha process to the energy given off in the proton-proton cycle. The key thing to bear in mind in terms of a stable phase (i.e. 'Main sequence') star like the Sun is that it is in pressure-gravity balance. The outer gas pressure balances the weight of its overlying layers. Any condition likely to disrupt this balance is therefore of paramount interest.

The stable lifetime of the Sun depends on how long before it consumes ninety percent of the hydrogen in its core. Theoretical investigations using data from nuclear reaction rates and cross sections suggest the Sun's Main Sequence lifetime at 8-10 billion years. Since it already has spent 4.5 billion of those years, there are anywhere from 3.5 to 5.5 billion years remaining.

Once the triple-alpha process gets underway and the energy balance declines, the Sun will have to compensate for the lost energy to sustain any kind of balance. Thus, the Sun's core must contract and convert gravitational potential energy into thermal energy. Meanwhile, ignition of hydrogen burning in the Sun's outer layers will create radiation pressure that forces the outer layers to expands. The Sun will then become a "*Red Giant*" and its new larger surface will be expected to engulf all the planets up to and including Mars.

Example Problem:

If the atomic mass of beryllium 8 (^{8}Be) = 8.00531u, verify that the first part of the triple-alpha fusion process is endothermic and has the value given.

Solution:

We have:

$Q = [\ 2(4.00260\ u) - 8.00531\ u]\ c^2$

$Q = [-\ 0.00011]\ 931.5\ MeV = 0.102465\ MeV = -\ 102.4\ keV$

Of course, not taken into account here is the gamma ray (γ) which also comes off. Hence we will have:

$(-102.4\ keV) + (E\ (\gamma)) = \mathbf{-95.7\ keV}$

So that:

$E\ (\gamma) = \mathbf{hc/\lambda = 6.7\ keV}$

Is the missing energy of the gamma ray photon, with the difference factored in yielding 95.7 keV.

Problems:

1) Calculate the wavelength of the gamma ray photon (in nm) which would be needed to balance the endothermic part of the triple –alpha fusion equation. (Recall here that $1\ eV = 1.6\ x\ 10^{-19}\ J$)

2) Verify the second part of the triple-alpha fusion reaction, especially the Q-value. Account for any differences in energy released by reference to the gamma ray photon coming off and specifically, give the wavelength of this photon required to validate the Q.

3) The luminosity or power of the Sun is measured to be $L = 3.9 \times 10^{26}$ watts. Use this to estimate the mass (in kilograms) of the Sun that is converted into energy every second. State any assumptions made and reasoning.

4) In a diffusion cloud chamber experiment, it is found that alpha particles issuing from decay of U238 ionize the gas inside the chamber such that 5×10^3 ion pairs are produced per millimeter and on average each alpha particle traverses 25 mm. Estimate the energy associated with each detected vapor trail in the chamber if each ion pair generates 5.2×10^{18} J.

5) Estimate the energy a solar proton would have to have in order to overcome the Coulomb barrier and undergo nuclear fusion with another proton.

APPENDIX V : Glossary

Absolute Magnitude: The apparent magnitude a star would have if observed from a standard distance of 10 parsecs (32.6 light years).

Altitude: The angular distance of a celestial object above the observer's horizon, measured along a vertical circle.

Aphelion: The position of a planet or other body in its orbit when farthest from the Sun or central mass. (For an artificial satellite it is usually called "apogee")

Apparent magnitude: A measure of the apparent brightness of a celestial object. This is gauged on a *magnitude scale* in which the brighter the object, the lower magnitude value, and vice versa. The scale is arranged such that each unit of magnitude corresponds to a difference in brightness factor or 2,512 times. Thus, five magnitudes difference is equivalent to a factor brightness difference (brightness ratio) of $(2.512)^5 = 100$ times.

Apparent Solar Time: Solar time corresponding to the true position of the Sun, e.g. as read by a sundial.

Astrolabe: One form of instrument capable of measuring both altitude and azimuth of celestial objects.

Astrology: The ancient pseudo-science which associates human destiny and human personalities with the locations and configurations of stars and planets in the sky. There is no foundation for it and it should not be confused with astrology.

Astronomical Unit: The mean distance between the Earth and the Sun, also known as the semi-major axis of the Earth's orbit. 1 AU = 1.496 x 10^{11} m.

Azimuth: The angle along the horizon measured eastward, from the north point to the intersection of the horizon with the vertical circle passing through the object.

Binary Star: A pair of stars which revolve around a common center of mass.

Black Body: A perfect absorber and perfect radiator of which stars are the best examples in the real world. All black bodies obey the Planck radiation law (see Planck function) and display what we call blackbody curves.

Cepheid Variable: a variable star named after the prototype, Delta Cephei. Type I Cepheids (classical), typically have periods from 2 – 100 days, and a range of median absolute magnitudes from (-1.5) to (-5). Characteristically, they are supergiants of spectral class F to G, The reason for brightness fluctuations can be attributed to regular pulsations of the star.

Comet: An object comprised of solid particles – mostly in the form of rocks and ice- as well as gases, which revolve around the Sun in highly eccentric orbits.

Conjunction: The apparent close approach of two heavenly bodies as detected by the eye in the night sky.

Constellation: An area of the sky named after a mythological character, animal or object. Its configuration is most often a purely coincidental "line of sight" effect with the actual members stars at widely differing distances.

Culmination: The time when a celestial object reaches the observer's meridian, marking its highest altitude.

Declination: The angular distance of a celestial object north or south of the Celestial Equator (defined as zero degrees Declination).

Distance modulus: A measure of the distance of a star defined as the difference between the star's apparent magnitude (m) and its absolute magnitude (M).

Diurnal Motion: The apparent daily rotation of the sky from east to west arising from the real rotation of the Earth from west to east.

Doppler shift: The apparent shift in the wavelength of a light source when the source is approaching or receding from an observer. If the source is approaching we perceive more vibrations per unit time, hence a higher frequency effect (or pitch, if a sound source). If the source is receding, there are fewer vibrations received per unit time, hence a lower frequency shift.

Eccentricity: For an ellipse or an elliptical orbit – the ratio of the distance between the foci to the major axis. Mathematically, it is defined:

$e = (a^2 - b^2)^{1/2} / a$

Where a is the semi-major axis and b is the semi-minor axis of the ellipse.

Ecliptic: The projection of the Earth's orbit on to the Celestial Sphere. (This is also defined as *the apparent yearly path of the Sun* against the background stars.)

Eddington-Barbier Relation: The important astrophysical approximation for a plane parallel atmosphere which states that the flux coming *out of the stellar surface* is equal to the source function at the *optical depth* $\tau = 2/3$. Thus:

$\pi(F_o) = 2\pi(I(\cos(\theta))) = \pi[a(\lambda) + 2(b(\lambda)/3]$

and $F_{\lambda o} = S(\lambda)\ \tau(\lambda) = 2/3$

Elongation: The apparent angular distance of a planet from the Sun as observed from Earth.

Equinox: One of two points (also called "equinoctial points") at which the Sun crosses the Celestial Equator in the course of a year. (The Vernal Equinox is on or about March 21st, and the Autumnal Equinox is on or about Sept. 22nd.) At either of these dates there is equal day and night all over the world.

Galaxy: A vast (e.g. 100-200 billion) assembly of stars gravitationally bound as a single physical system.

Light Year: The DISTANCE light travels in one year, equal to roughly 5, 880,000,000 miles (Multipl7 186,300 miles per second time the number of seconds in a year).

Lunation: The time interval between successive New Moons, or about 29 ½ days, also equal to the synodic period of the Moon.

Mean Solar Time: A uniform solar time obtained by dividing the year into 365 days of equal length. It may differ by as much as 16 minutes from apparent solar time.

Meridian: The great circle for longitude, or sidereal time, or Right Ascension, passing directly through the observer's zenith.

Meteor: The luminous streak observed when a particle enters the Earth's atmosphere.

Meteorite: The portion of a particle or object that survives passage through the Earth's atmosphere and strikes the ground.

Meteoroid: The particle or object moving in space before any encounter with Earth's atmosphere.

Obliquity of the Ecliptic: The angle between the plane of the Earth's equator and the plane of its orbit (ecliptic), or 23 ½ degrees.

Opposition: The position of a superior planet (i.e. beyond the Earth's orbit) when it lies on the extension of the line joining the Earth and the Sun. The planet is therefore diametrically opposite to the Sun in the sky and rises when it sets.

Occultation: The eclipsing of one celestial body by another in the line-of-sight.

Period-Luminosity Law: The empirical law, discovered by Henrietta Leavitt, which relates the periods of Cepheid variables to their intrinsic brightness. The more intrinsically bright the star, the longer its period.

Phase: The appearance or form of that part of the disk of the Moon, or inferior (inner to the Earth's orbit) planet, which is illuminated by the Sun as seen from Earth.

Planck function: The function which describes the distribution of radiation for a black body, expressed:

$B(\lambda) = \{(2 hc^2)/ \lambda^5\} [1/ \exp(hc/\lambda kT) - 1)]$

where h is Planck's constant, c is the speed of light, T is the absolute temperature, k is the Boltzmann constant, and λ defines the wavelength.

Retrograde Motion: The apparent westward motion of a planet on the celestial sphere.

Russell-Vogt Theorem: A theorem which states that the mass and chemical composition of a star determine its entire structure provided it derives its energy from nuclear reactions.

Schwarzschild Radius: The critical radius beyond which a given stellar mass will be too dense to permit light to escape – hence the mass becomes a black hole.

Sidereal Time: A local time based on the rotation of the celestial sphere. The sidereal day of 24 sidereal hours begins when the vernal equinox (First point of Aries) crosses the observer's meridian (upper transit).

Solar Constant: The mean intensity of solar radiation received by a unit area per unit time just outside the Earth's atmosphere. This can also be obtained for any star, as the value S in the equation:

$\pi F = S (r/R)^2$

Where $(r/R) = 1/\alpha$

The inverse (or reciprocal) of the angular radius of the star.

Spectral sequence: The sequence of spectral classes for stars (O, B, A, F, G, K and M) arranged in order of decreasing temperatures. Thus, type 'O' stars are the hottest and type 'M' the coolest.

Vertical circle: Any great circle passing through the zenith.

Zenith: The point on the celestial sphere that is located directly above an observer, i.e. at an altitude of 90° above his horizon.

Selected Solutions to Extra Problems

Chapter II: Problem (1)

Jacksonville, Florida is located at approximately latitude 30° 20' N.

a) Sketch a declination diagram for this location and carefully identify the position of the North Celestial Pole and the approximate altitude of the Pole star.

b) If the star Procyon is visible and has a declination of approximately $\delta = +5°$ then find its maximum altitude from this location and its zenith distance, z.

Solution:

The declination diagram for Jacksonville is shown:

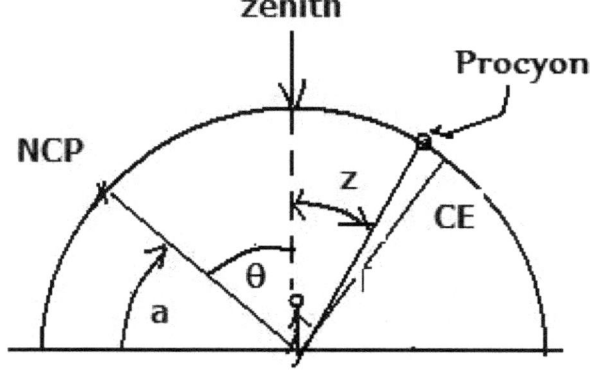

For Jacksonville, FLA.:

$a = 30° \; 20'$ (altitude of NCP)

Here, a also denotes the approximate altitude of the Pole star, or a = 30° 20'.

b) The maximum altitude first depends on getting the zenith distance z, since a_{max} = (90° - z).

We can work all of this out from the geometry. We note first that a = 30° 20'. Then:

θ = 90° − a = 90° - 30° 20' = 59° 40'

Since we know that δ = +5° for Procyon, and:

90° = θ + z + δ, then:

z = (90° - θ - δ)

z = (90° - 59° 40' - 5°) = (90° - 64° 40')

So: z = 25° 20' (Ans.)

Therefore: a_{max} = 90° - 25° 20' = 64° 40'

Problem (3):

How would the values obtained in (2) for Procyon (and Jacksonville) change for a location in Barbados?

Solution:

Barbados is at latitude 13° therefore the approximate altitude of the Pole star, a = 13°. Then:

θ = 90° − a = 90° - 13° = 77°

And: $z = (90° - \theta - \delta) = (90° - 77° - 5°) = 8°$

Therefore: $a_{max} = 90° - 8° = 82°$ (Ans.)

Chapter V: Problem (1)

The star Acrux (*alpha Crucis*) at RA = 12h 21m, is observed to have a local hour angle = 30 deg on May 10th for a given location.

(a) What is *the local sidereal time.*

(b) At what local mean time and standard time would Acrux transit?

(c) What is the approximate LST at noon on the same date?

Solution:

a) LHA = LST - RA And we know: LHA = 30° or:

LHA = (30°) / (15°/h) = 2h 00m

So:

LST = LHA + RA = 2h 00m + 12h 21m = 14h 21 m

Ans. 14h 21 m

LTT (local mean time of transit)= Star's RA - Sun's RA

May 10th is 41 days before June 20, the summer solstice (assume non-leap year) so:

Sun's RA = 6h - 41 days x (4 min/day) = 0h - 164 mins
= 6 h - 2h 44 m

Sun's RA = 3 h 16 m

LTT = 12h 21m - 3h 16 m = 9 h 05 m

Or: 12 h 00m + 9 h 05 m = 9h 05 m

c) From (a) the LST on the date is 14h 21 m. This is 14h 21 m past noon. Since noon is 14h 21 m *earlier*, then:

LST(noon) = 12h 21m - 14h 21 m = -2h 00m, or:

24h 00m - 2h 00m = 22h 00m.

Chapter V: Problem (3):

You are required to calculate the hour angle of the Sun (HA ☉) from a place with longitude 163 ° 14' E. The observation is made at a standard time of 8:46 a.m. on March 10, the standard time being referred to the meridian of longitude 165 ° E. (The Equation of Time for the particular date is approximately E = - 10 min. Note that Equation of Time is just the difference (Apparent time − Mean time).

Solution:

Given the longitude circle is 1 ° 46' west of the standard longitude this means the time difference is about: 4 min. + 3 mins. = 7 mins. earlier or 8:39 a.m. LMT. Then the Equation of Time for the particular date is approximately E = - 10 min. so:

App. Solar time = 8:39 a.m. − 10 mins. = 8:29 a.m.

Since the hour angle of the Sun (HA ☉) is referenced with respect to local noon then this means we have:

(HA ☉) = 90 ° − (45 ° + 7 ¼ °) = 90 ° − 52 ¼ °

(HA ☉) = 38 ° E., approximately!

Chapter XII: Problem (1):

You are located in Miami, Florida and the sidereal time at 9.30 p.m. local time for your location on this date, is 9 h 13 m, approximately. Saturn is visible and is at 13 h 44 m Right Ascension, and at a Declination of (- 7° 56').

If your latitude is 25.°75 north, find Saturn's position in terms of its altitude and azimuth. *Use both* the astronomical triangle and matrix method to obtain your answer.

Solution (I): *Astronomical triangle:*

The sidereal time (ST) is given as: 9 h 13 m

Next, we change the Right Ascension (13 h 44 m) to hour angle using:

h = ST - RA

So: h = 9 h 13 m − 13 h 44 m = - 3h 31m

This is then converted into degrees, using the fact that

there are 15 degrees/ hr.

So: -3 h 31 m ≈ -52°.5

We find the zenith distance, z, using an astronomical triangle using sides with hour angle, declination and latitude. This implies the spherical version of the law of cosines, and by analogy with the previous solution we have:

$\cos z = \sin(\delta) \sin(\text{Lat}) + \cos(\delta) \cos h \cos(\text{Lat})$

We note the following respective values:

$\sin(\delta) = \sin(-7° 56') = \sin(-7.°93) = -0.138$

$\sin(\text{Lat}) = \sin(25.°75) = 0.434$

$\sin h = \sin(-52°.5) = -0.793$

$\cos(\delta) = \cos(-7.°93) = 0.990$

$\cos h = \cos(-52°.5) = 0.608$

$\cos(\text{Lat}) = \cos(25.°75) = 0.900$

so, effectively:

$\cos z = (-0.138)(0.434) + (0.990)(0.608)(0.900)$

$\cos z = -0.0598 + 0.541 = 0.481$

$z = \arccos(0.481) = 61.°1$

Then,

a (altitude) = $90° - 61.°1 = 28.°9$

But which direction?

This requires the azimuth, A:

The appropriate astronomic triangle yields the following equation for A, azimuth:

$\tan A = -\cos(\delta) \sin h / [\sin(\delta)\cos(Lat) - \cos(\delta) \cos h \sin(Lat)]$

Take the numerator first and compute it:

$-(0.990)(-0.793) = 0.785$

Then the denominator:

$= (-0.138)(0.900) - (0.990)(0.608)(0.434)$

$= -0.124 - 0.261 = -0.385$

so that:

$\tan(A) = (0.785)/(-0.385)$

$\tan(A) = -2.03$

$A = \arctan(-2.03) = -63°.8$

This is positive, so this angle is subtracted from $360°$ to get:

360° - 63°.8 = 296°.2 (Or slightly north, e.g. 26°.2, of due east)

Solution (II): Matrix Method

We need to perform the matrix operations in the specific order:

(x)
(y)
(z) A,a = R3(-180°) R2(90 - lat.) (XYZ(h, δ))

Where:

$$R_3(180) := \begin{pmatrix} \cos(180) & \sin(180) & 0 \\ -\sin(180) & \cos(180) & 0 \\ 0 & 0 & 0 \end{pmatrix}$$

Therefore: R3(-180) =

(-1   0   0)
(0   -1   0)
(0   0   1)

And: R2(90° - lat.) =

(sin lat.   0   - cos lat.)
(0   1   0)
(cos lat.   0.   sin lat.)

And for which we have:

sin (lat.) = sin (25.°75) = 0.434

cos (lat.) = cos (25.°75) = 0.900

Thence, $R_2(90° - lat.)$ =

$$\begin{pmatrix} 0.434 & 0 & -0.900 \\ 0 & 1 & 0 \\ 0.900 & 0 & 0.434 \end{pmatrix}$$

Finally:

$$\begin{pmatrix} x \\ y \\ z \end{pmatrix} h, \delta = \begin{pmatrix} \cos \delta & \cos h \\ \cos \delta & \sin h \\ \sin \delta & - \end{pmatrix}$$

where:

$\sin (\delta) = \sin (-7.°93) = -0.138$

$\cos (\delta) = \cos (-7.°93) = 0.990$

$\cos h = \cos (-52°.5) = 0.608$

$\sin h = \sin (-52°.5) = -0.793$

Therefore:

(0.990   0.608)
(0.990   -0.793)
(-0.138 ……….) =

(0.601)
(-0.785)
(-0.138)

Whence:

R3(-180°) R2(90° - lat.) (XYZ(h, δ)) =

(-0.385)
(0.785)
(0.481)

The last element in the column yields the altitude, so:

a = arc sin(0.481) and a = 28.°75

Meanwhile, the azimuth A =

arc tan (y/x) = arc tan (0.785/ -0.385) = -2.03

Therefore: A = arc tan(-2.03) = -63.°8

And, since its' negative, we *must subtract* from 360 degrees:

A= 360° - 63.°8 = 296.°2

Chapter XV: Problem (2):

In the atmosphere of a star at a particular layer, the temperature is 5650 K and there are approximately 1.45 x 10^{19} free electrons. If the total gas pressure is 8.3 x 10^3 Pa find: a) the electron pressure, and b) the total number of neutral and ionized atoms.

Solution:

We know:

$P = (N_o + N_e) k T$

Where N_o is the number of atoms of all kinds, both neutral (unionized) and ionized and N_e is the number of free electrons (the greater this number, the higher the degree of ionization). The total gas pressure therefore is the sum of the two pressures:

$P = N_o k T + N_e k T$

Then: $N_e k T$ = (1.45 x 10^{19})(1.38 x 10^{-23} J/K) (5650 K) = 1.13 Pa

But P, the total gas pressure is 8.3 x 10^3 Pa, so:

P = 8.3 x 10^3 Pa = $N_o k T + N_e k T = N_o k T$ + 1.13 Pa

Then: $N_o k T$ = 8.3 x 10^3 Pa - 1.13 Pa ≈ 8.3 x 10^3 Pa

So: N_o = (8.3 x 10^3 Pa) / k T = (8.3 x 10^3 Pa) / (1.38 x 10^{-23} J/K) (5650 K)

N_o = 1.06 x 10^{23}

Where N_o is the total number of neutral and ionized atoms.

(3) Use the temperature obtained in the last sample problem above to obtain an estimate of the rate of energy generation in the solar core. Assume an 80% hydrogen content in the core. Also use the approximate equation for sample worked problem (2) to obtain an estimate of the solar luminosity. By what amount is this in error if the actual luminosity is $L = 3.9 \times 10^{26}$ W.

Solution:

The temperature for the last sample problem (p. 190) was given as $T = 1.2 \times 10^7$ K. From this we estimate the rate of energy generation in the solar core from:

$\varepsilon = 2.5 \times 10^6 \, (\rho \, X^2) \cdot (10^6/T)^{2/3} \exp[-33.8(10^6/T)^{1/3}]$

Where $X = 0.80$ (80% hydrogen content)

Then:
$\varepsilon = 2.5 \times 10^6 \, (1.4 \text{ gcm}^{-3} \times 0.80^2) \cdot (10^6/1.2 \times 10^7 \text{ K})^{2/3} \exp[-33.8(10^6/1.2 \times 10^7 \text{ K})^{1/3}]$

Note that the density $\rho = 1.4$ gcm^{-3} since cgs units are needed for this formula!

Then:

$\varepsilon = 2.5 \times 10^6 \, (0.896) \cdot (0.1837) \cdot (5.1) = 2.1 \times 10^6$

This is in ergs per gram per sec so we convert back to SI units using the fact that 1 erg = 10^{-7} J and 1000 g = 1 kg. Then with these conversions:

$\varepsilon = 2.1 \times 10^2$ J kg^{-1} s^{-1}

We estimate the luminosity using:

$dL/dr = \varepsilon (4\pi r^2 \rho)$

So: $L/r \approx \varepsilon (4\pi r^2 \rho) = 4\pi (2.1 \times 10^2$ J kg^{-1} s$^{-1})(7 \times 10^8$ m$)^2 (1400$ kg m$^{-3})$

$L \approx 4\pi (2.1 \times 10^2$ J kg^{-1} s$^{-1})(7 \times 10^8$ m$)^3 (1400$ kg m$^{-3})$
$= 1.26 \times 10^{33}$ J/s

This is different from the actual solar luminosity of:

$L_\odot = 3.9 \times 10^{33}$ J/s

By only about 0.2 order of magnitude.

Clearly, expediently turning a derivative into a division is enough to give a reasonable approximation, if not exact result!

Chapter XVII: Problem (1):

A star has a gray atmosphere for which the Eddington approximation: $T^4 = \tfrac{3}{4} T_e^4 (\tau + 2/3)$ is valid, where T_e denotes the effective temperature. Use this approximation to obtain the fraction of outward intensity escaping from the star's surface.

Solution:

We begin by assuming: a) radiative equilibrium, and b) "LTE" or local thermodynamic equilibrium.

First, find the intensity radially coming out in all directions, from:

$$I(\Theta=0) = \int_0^\infty B(T)\exp(-\tau)\,d\tau$$

$$= \sigma/\pi \int_0^\infty (T)^4 \exp(-\tau)\,d\tau$$

$$I(\Theta=0) = 3\sigma/4\pi\, T_e^4 \int_0^\infty (\tau + 2/3)\exp(-\tau)\,d\tau$$

$$\therefore I(\Theta=0) = 5\sigma/4\pi\, (T_e^4)$$

Now, the part of the above which originates above the critical layer- call it layer τ_0 is:

$$I(\tau < \tau_0) = 3\sigma/4\pi\, T_e^4 \int_0^\infty (\tau + 2/3)\exp(-\tau)\,d\tau$$

$$I(\tau < \tau_0) = 3\sigma/4\pi\, T_e^4\, [5/3 - (\tau_0 + 2/3)\exp(-\tau_0)]$$

The fraction originating above any given optical depth is then:

$$f(I(\tau < \tau_0)) = 1 - (0.6\tau_0 + 1)\exp(\tau_0)$$

The fraction escaping from the star's surface is then that for which $\tau_0 = 0$, so:

$$1 - (0.6\,\tau_0 + 1)\exp(\tau_0) = 1 - (0.6(0) + 1)\exp(0) =$$
$$1 - (0 + 1)(1) = 1 - 1 = 0$$

Chapter XVII: Problem (2):

The star Suhail has a (B − V) color index of +1.7. Use this and the Table on p. 140 to obtain the net flux (H) passing through the Suhail's surface. How might you estimate the intensity I from this and the mean intensity J?

Solution:

From the (B − V) Table on p. 140 we see that a (B − V) color index of +1.7 corresponds to $\log T_e = 3.52$

Then: antilog (3.52) = 3 300 K

which is the effective temp. (T_e)

Then the net flux H passing through Suhail's surface is obtained from:

$\sigma T_o^4 / \pi = 2H$

where the *boundary temperature*

$T_o = T_e / 1.189 = 3300 \text{ K} / 1.189 = 2770 \text{ K}$

so:

$H = \sigma T_o^4 / 2\pi$

$= [(5.67 \times 10^{-8} \text{ W m}^{-2} \text{ K}^{-4})(2770 \text{K})^4 / 2\pi$

$H = 5.3 \times 10^5$ W

The intensity I and the mean intensity J can be estimate from the above using :

a) $J = 2H/3 = 2(5.3 \times 10^5 W) / 3 = 3.5 \times 10^5 W$

and

b) $I_1(\tau) = 3H\tau$ (where I_1 is *the forward flux*)

and setting $\tau = 2/3$:

$I_1(\tau) = 3H(2/3) = 2H = 2(5.3 \times 10^5 W) = 1.1 \times 10^6 W$

Chapter XVII: Problem (3):

For many stars, the solar constant S can be computed if its angular diameter is known. If the angular radius of a star is: $\alpha = (R/r)$ with r the distance to Earth and R the star's linear radius then:

$\pi F = S(r/R)^2$

(Note: that α is measured in *radians*)

If the Sun's angular radius is 959.63 arcsec then find the solar constant S. (Hint: you can use the solar flux πF already computed from sample problem 1).

Solution:

We re-arrange to find:

$S = \pi F / (r/R)^2$

Where we already know that the flux:

$\pi F = 6.3 \times 10^7 \, Jm^{-2} \, s^{-1}$

$\alpha = 959.63$ " but this must be in radians before using the equation.

One radian = 57.3 degrees

= 57.3 deg/rad x (3600"/ deg)= 206 280 "

Then:

$\alpha = (R/r) = 959.63"/ 206\,280"/rad = 0.00465 \, rad$

So: $(r/R) = 1/0.00465 \, rad = 215 \, rad^{-1}$

Therefore:

$S = [6.3 \times 10^7 \, Jm^{-2} \, s^{-1}]/ [215 \, rad^{-1}]^2 = 1362 \, W/m^2$

Chapter XVII: Problem (4):

Find the solar constant S for α Lyrae (Vega) if we know (from Hanbury and Brown's measurements) that its angular diameter is 0.0032 arcsec, and it has a (B- V) color index of 0.00. Show all working and state any assumptions.

Solution:

We use: $\pi F = \sigma (T_{eff})^4$

And get T_{eff} from the table on page 140, where we find a (B – V) index = 0.00 corresponds to:

$\log T_{eff} = 4.03$

Then antilog (4.03) = 10 700 so T_{eff} = 10 700 K

$\pi F = \sigma (T_{eff})^4 =$

$(5.67 \times 10^{-8}$ W m^{-2} K^{-4}) $(10\ 700$ K$)^4 = 7.4 \times 10^8$ Jm^{-2} s^{-1}

The solar constant for Vega is obtained from:

$S = \pi F / (r/R)^2$

and we know: $\alpha = (R/r) = 0.0032$ arcsec

Or, $\alpha = 0.0032$ "/ 206 280"/rad $= 1.55 \times 10^{-8}$ rad

$(r/R) = 1/1.55 \times 10^{-8}$ rad $= 6.4 \times 10^7$ rad^{-1}

Therefore the solar constant for Vega is:

$S = [7.4 \times 10^8$ Jm^{-2} s^{-1}/ $[6.4 \times 10^7$ rad$^{-1}]^2$

$= 1.7 \times 10^{-7}$ W/m^2

<u>Chapter XVIII: Problem (1):</u>

It is said that the Earth is *"enveloped in high temperature coronal material"*. Verify this by obtaining the temperature of coronal material at the Earth's distance from the Sun: $d = 1.5 \times 10^{11}$ m.

Solution:

We use: $T(r) = T_o (R_o / r)^{2/7}$

Where d = r and $T_o = 2 \times 10^6$ K

Then: $R_o / r = (7 \times 10^8 \text{ m} / 1.5 \times 10^{11} \text{ m}) = 0.0046$

Then:

$T(r) = 2 \times 10^6 \text{ K} (0.0046)^{2/7} = 4.3 \times 10^5$ K

Or, 430 thousand degrees Kelvin, so definitely high temperature coronal material.

Chapter XVIII: Problem (2):

Find the coronal pressure at the Earth's distance if that pressure as a function of distance r is given by:

$p(r) = p_o \exp\{-7R_o[1 - (R_o/r)^{5/7}]/5H\}$

In cgs units, $H \approx 10^{10}$ m is known as the "scale height". (Take $p_o = 0.2$ dynes/cm²)

Solution:

$p_o = 0.2$ dynes/cm²

Then:

$p(r) = (0.2) \exp\{-7R_o[1 - (0.0046)^{5/7}]/5 \times 10^{10} \text{ cm}\}$

$= 1.1 \times 10^{-5}$ dynes/cm² or 1.1×10^{-6} Pa

Chapter XVIII: Problem (4):

According to Mihalas (*op. cit.*, p. 525) the coronal "flow velocity" v can be deduced from the flux F based on the equation:

$$F = 4 \pi r^2 n (v)$$

where n denotes the particle density, i.e. per unit volume. Compute this velocity: a) at the distance of the Earth (See: #1) and b) at the distance associated with the coronal temperature from Problem #2.

You may use the equation (Mihalas, p. 524):

$$n(r) = n_o (r/R_o)^{2/7} \exp\{- 7R_o[1 - (R_o /r)^{5/7}]/ 5H\}$$

taking $n_o = 4 \times 10^8 /cm^3$

Solution:

Again, straight c.g.s. units can be used with the equation, so $R_o = 7 \times 10^{10}$ m, and the distance of Earth to Sun is $r = 1.5 \times 10^{13}$ cm.

Then, writing the form out:

$$n(r) = (4 \times 10^8 /m^3) (214)^{2/7} \exp\{- 7R_o[1 - (0.0046)^{5/7}]/ 5(10^{10} cm))$$

$$n(r) = 1.2 \times 10^6 /m^3$$

The flux can be computed using S.I. units:

$$F = 2/7 \ [4 \pi (1.5 \times 10^{11} m) (1.1 \times 10^3 W \ m^{-1} K^{-1}) (4.3 \times 10^5 K \ )$$

409

Where the last two factors are the values for the kinetic temperature arising from the corona (T) and the conductivity κ, at the Earth's distance. Then:

F = 2.55 x 10^{20} W = 2.55 x 10^{20} Joules/ sec

But we understand that a conversion factor of 10^7 is needed to render it consistent with earlier c.g.s units. Then:

F = 2.55 x 10^{27} ergs/s

The velocity is then: v = F / 4 π r^2 n

= (2.55 x 10^{27}ergs/s)/ 4 π (1.5 x 10^{11} m)^2 4 x 10^{14} /m^3
≈ 0

At 2 R$_o$ we find the same results.

Chapter XIX: Problem (3):

There are 3 conditions for inner Lindblad resonance in a spiral galaxy:

2(Ω_p - Ω) / κ$_o$ = -1, 0 and +1

Find the values of Ω_p which satisfy each.

Solution:

We take the simplest first:

2(Ω_p - Ω) / κ$_o$ = 0

Then:

$2(\Omega_p - \Omega) = 0$

Or: $\Omega_p = \Omega/2$

At the sun's distance, $\Omega \approx 10^{-15}$ rad s^{-1}

So: $\Omega_p = (10^{-15}$ rad s$^{-1})/2$

Next, let: $2(\Omega_p - \Omega)/\kappa_o = -1$

Then: $2(\Omega_p - \Omega) = -\kappa_o$

$\Omega_p = -\kappa_o/2 + \Omega$

Where:

$\kappa_o = 1.35 \ \Omega = 1.35 (10^{-15}$ rad s$^{-1}) = 1.35 \times 10^{-15}$ rad s^{-1}

So:

$\Omega_p = 10^{-15}$ rad s$^{-1} - (1.35 \times 10^{-15}$ rad s$^{-1})/2$

$\Omega_p \approx 0.33 \times 10^{-15}$ rad s$^{-1} \approx 3.3 \times 10^{-16}$ rad s^{-1}

Finally: let: $2(\Omega_p - \Omega)/\kappa_o = +1$

Then: $2(\Omega_p - \Omega) = +\kappa_o$

$\Omega_p = \kappa_o/2 + \Omega = 1.35 \times 10^{-15}$ rad s$^{-1}/2 + 10^{-15}$ rad s^{-1}

$\Omega_p \approx 1.67 \times 10^{-15}$ rad s^{-1}

Chapter XIX: Problem (4):

The Andromeda Galaxy has a mass of approximately 1.2×10^{12} solar masses and a radius of 22,000 pc. For a Sun-sized star, one-third of the way to the Andromeda rim, compute: i) the potential V(r), ii) the period of this star, iii) the epicyclic frequency (κ) if the ratio χ =2, iv) the specific angular momentum J.

Solution:

i) The potential V(r) is defined: V(r) = - GMm/r

Where m = 2×10^{30} kg (for sun-sized star)

And: M = $(1.2 \times 10^{12})(2 \times 10^{30}$ kg) = 2.4×10^{42} kg

The value of r (distance to sun-sized star):

= 2/3 (22,000 pc) = 44,000 pc/3 = 14, 600 pc

Then in meters:

r = $(1.46 \times 10^4$ pc) (3.26 Ly/pc)(9.46×10^{15} m/Ly)

r = 4.5×10^{20} m

V(r) =

- $(6.7 \times 10^{-11}$ N-m^2/kg^2)(2.4×10^{42} kg)(2×10^{30} kg) /r

V(r) = $-7.1\ 10^{44}$ J

(ii)The period is found from Kepler's 3rd law:

$(M+ m)(P1/ P2)^2 = k(a1/ a2)^3$

Or in simplified form, with Earth values: a2 = 1 AU, P2 = 1 yr:

$(M+ m)(P1)^2 = (a1)^3$

Where: a1 = a = (1.46 x 10^4 pc) (2 x 10^5 AU /pc) =

2.92 x 10^9 AU

And P1 = $[(a1)^3 / (M+ m)]^{1/2}$

P1 = $[(2.92 \times 10^9 \text{ AU})^3/ 1.2 \times 10^{12}]^{1/2}$

And note the denominator must be the total solar mass value of the Andromeda galaxy, since we may neglect m owing to the fact: m < < M.

P = 1.4 x 10^8 yrs

(iii) The epicycle frequency is $\kappa_0 = \chi \Omega$

Where χ is the number of of epicycle oscillations per orbit about a galactic center. For this case we are given that: $\chi = 2$, so that:

$\kappa_0 = 2 \Omega$

Where: $\Omega = (2 \pi \text{ rad})/ T(s)$

And T(s) =

(1.4 x 10^8 yrs) (365.25 days/yr.) (86,400 s/ day)

T(s) = 4.5 x 10^{15} s

Ω = (2 π rad)/ (4.5 x 10^{15} s) = 1.4 x 10^{-15} rad s^{-1}

The specific angular momentum for a star located in a galactic disk is:

p$_\varphi$ = J = r^2 Ω (r)

Therefore:

J = (4.5 x 10^{20} m)^2 (1.4 x 10^{-15} rad s^{-1})

J = 6.2 x 10^5 m^2 s^{-1}

Chapter XX: Problem (1):
Examine the magnetogram below for a sunspot region.

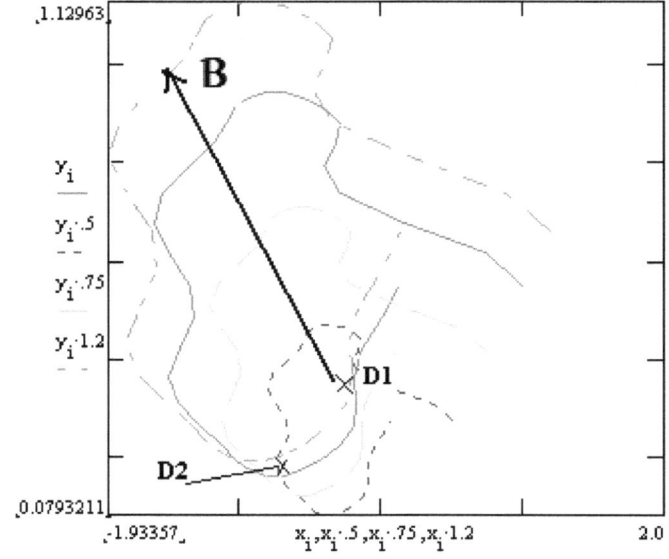

Each increment along "Delta Y" or "Delta X" denotes a change of 250 km while each contour difference is a separation of 250 G. The B-vector direction has been indicated as shown.

From this information and the contour map work out the gradients $\partial B(x)/\partial y$ and $\partial B(y)/\partial x$.

Then: Obtain the current helicity density, and the force-free scale factor (α) applicable to the region.

Solution:

We first note that $\partial B(x)$ effectively crosses four contours, or (4 x 250 G) = 1000 G. The separation dy (or ∂y) from the vertical axis amounts to $\approx$ 3.2 unit(s) or 3.2 x 250 km $\approx$ 800 km. Thus:

$\partial B(x)/\partial y$ = (1000 G)/ 800 km = 1.25 G/ km

In a similar way, we find for $\partial B(y)/\partial x$:

$\partial B(y)/\partial x \approx$ (1000 G) / (1.4 x 250 km) $\approx$

1000 G/ 350 km $\approx$ 2.8 G/ km

Then the current helicity density is:

$H_z(c)$ = [$\partial B(x)/\partial y$ - $\partial B(y)/\partial x$] B_z

$\approx$ [1.25 G/ km - 2.8 G/ km] 350 G

$H_z(c) \approx$ -542 G^2/ km

and the negative sign indicates that *the force-free parameter* α is also negative, and will have magnitude:

$$\alpha \approx -[\partial B(x)/\partial y - \partial B(y)/\partial x] / B_z$$

$$\approx -[1.25\ G/km - 2.8\ G/km] / 350\ G$$

$$\alpha \approx -(1.55 \times 10^{-3}\ m^{-1})\ G / 350\ G \approx -4.4 \times 10^{-6}\ m^{-1}$$

By "*hemispheric helicity rule*" the sign fixes the region in the Northern solar hemisphere.

Chapter XX: Problem (2):

Given the force-free scale factor obtained in (1) and a vertical current density $J_z \approx 0.012$ A m^{-2} estimate the magnitude of the magnetic component B_z.

Solution:

We know: $\alpha = \mu_0 J_z / B_z$

Then: $B_z = \mu_0 J_z / \alpha$

Where: $\mu_0 = 4\pi \times 10^{-7}$ H/m

Solving for the magnetic field component:

$B_z = (4\pi \times 10^{-7}\ H/m)(0.012\ A\ m^{-2}) / -4.4 \times 10^{-6}\ m^{-1}$

$B_z = -3.4 \times 10^{-3}$ T

Chapter XXI: Problem (3):

A Calcium line in the spectrum of α Centauri has a wavelength of 3968.20 Å. The same line in the solar spectrum has a measured wavelength of 3968.49 Å.

Find the radial velocity of α Centauri relative to the solar system. Is it approaching or receding?

Solution:

The normal position of the line is that in the solar spectrum since there is no appreciable radial velocity of the Sun with respect to Earth. Thus:

$\Delta\lambda = (\lambda - \lambda_o)$

Where: $\lambda_o = 3968.49$ Å and $\lambda = 3968.20$ Å

Therefore:

$(\lambda - \lambda_o) = (3968.20$ Å $- 3968.49$ Å$) = -0.29$ Å

The negative sign denotes motion toward the observer. Hence, approaching.

Since: $(\Delta\lambda / \lambda) = v/c$

The velocity $v = c (\Delta\lambda / \lambda) = c (-0.29$ Å$/ 3968.49$ Å$)$

$c = 3 \times 10^8$ ms^{-1}

so: $v = 3 \times 10^8$ ms^{-1} $(7.3 \times 10^{-5}) = 21\,900$ ms^{-1}

Chapter XII: Problem (2):

A rocket ship of length 100m travels at v/c = 0.6. It carries a radio receiver in its nose. A radio pulse is emitted from a stationary space station just as the ship passes by.

a) How far from the space station is the nose of the rocket at the instant the radio signal arrives at the nose?

b) By space station time, what is the time interval between the arrival of this signal and its emission from the station?

c) What is the time interval determined from measurements in the rocket ship's rest frame?

Solution:
It helps here to draw a sketch of the situation:

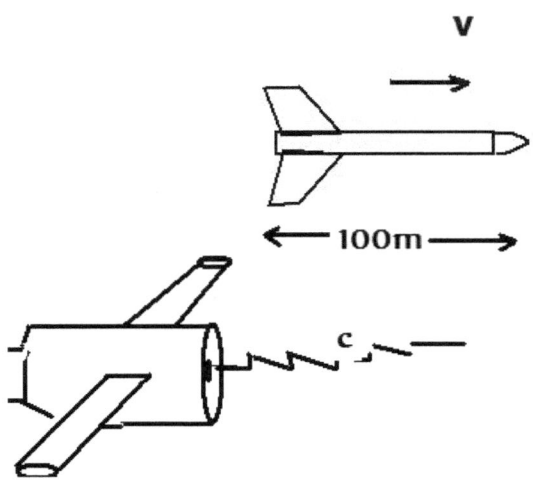

The foregoing diagram basically takes the key aspects into account, including the length of the rocket and its relation to the space station with signal dispatched.

a) The key thing to note here is that the distance covered (before the radio signal arrives at the nose) must take into account not only the length of the rocket (L = 100 m) but also the time it takes to travel the length L of the rocket at the speed of light.

Thus the distance D of the space station to the nose of the rocket is:

$D = L + d$

Where: $d = ct$

To obtain t: $t = 100m/3 \times 10^8 \, ms^{-1} = 3.3 \times 10^{-7} \, s$

Then: $d = (3 \times 10^8 \, ms)(3.3 \times 10^{-7} \, s) = 100m$

So: $D = L + d = 100 \, m + 100 \, m = 200 \, m$

b) The time required in the space station's rest frame must be based on the total distance covered in (a). Thus the time t' is:

$t = 200m/ 3 \times 10^8 \, ms^{-1} = 6.7 \times 10^{-7} \, s$

The time in the rocket's rest frame is just based on the distance its nose is from the station at signal reception or L' = 100m.

Then: $t' = 100m/3 \times 10^8 \, ms^{-1} = 3.3 \times 10^{-7} \, s$

Chapter XXIII: Problem (1):

Observations on the quasar 3C-9 indicate that when it emitted the light that just reached Earth it was receding at a velocity of 0.8c. One of the lines identified in its spectrum has a wavelength of 1200 Å when emitted from a stationary source. At what wavelength must this line have appeared in the spectrum of 3C-9?

Solution:

The velocity of recession is given as v = 0.8 c. The shift in spectral lines expected is:

$\Delta\lambda = (\lambda - \lambda_o) = (\lambda - 1200$ Å$)$

Where $\lambda_o = 1200$ Å is the stationary line position (e.g. from a stationary source). Then we need to solve for λ, given:

$(\Delta\lambda/\lambda) = v/c = (\lambda - 1200$ Å$)/\lambda = 0.8$

Using basic algebra:

$(\lambda - 1200$ Å$) = 0.8 \lambda$

Or:

$\lambda - 0.8 \lambda = 1200$ Å

$0.2 \lambda = 1200$ Å or $\lambda = 1200$ Å$/0.2$

$\lambda = 6000$ Å